A MOUSE IN A CAGE

A Mouse in a Cage

*Rethinking Humanitarianism and
the Rights of Lab Animals*

Carrie Friese

NEW YORK UNIVERSITY PRESS
New York

NEW YORK UNIVERSITY PRESS
New York
www.nyupress.org

© 2025 by New York University

This work is licensed under the Creative Commons Attribution-NonCommercial-NoDerivs 4.0 license (CC BY-NC-ND 4.0). To view a copy of the license, visit https://creativecommons.org/licenses/by-nc-nd/4.0.

Please contact the Library of Congress for Cataloging-in-Publication data.

ISBN: 9781479833474 (hardback)
ISBN: 9781479833481 (paperback)
ISBN: 9781479833511 (library ebook)
ISBN: 9781479833504 (consumer ebook)

New York University Press books are printed on acid-free paper, and their binding materials are chosen for strength and durability. We strive to use environmentally responsible suppliers and materials to the greatest extent possible in publishing our books.

The manufacturer's authorized representative in the EU for product safety is Mare Nostrum Group B.V., Mauritskade 21D, 1091 GC Amsterdam, The Netherlands. Email: gpsr@mare-nostrum.co.uk.

Manufactured in the United States of America

10 9 8 7 6 5 4 3 2 1

Also available as an ebook

For my dad, Walter J. Friese III

1949–2021

And for my mentor, Adele E. Clarke

1945–2024

CONTENTS

Introduction   1

1. Suffering   19
2. Care   39
3. Killing   55
4. Sacrifice   74
5. Compassion   92
6. Consent   108

Conclusion   127

Acknowledgments   145

Appendix: On Methodology   149

Notes   159

Bibliography   181

Index   201

About the Author   209

Introduction

I am spending a day shadowing Janet, an animal technician who has been working with laboratory animals for roughly 35 years. After changing into issued scrubs and taking an air shower, I meet Janet in the part of the animal facility containing the aged-mouse colony. As I put on my hairnet, gloves, and face mask, Janet gives me a brief history of the mice she will be working with today and whom I will be watching. Janet explains that the mice she looks after are for a laboratory that is interested in questions about the relationships between aging and immunity, particularly in regard to vaccine development. She tells me that the BALB/c mouse is the primary strain used as a model organism in immunological research. The problem is that males of this strain fight—even with littermates. Aged, male BALB/c mice, as a result, must be housed in a cage alone. Housing mice for up to three years alone in a cage is not something the Institute considers ethical, since it's thought that mice need community and companionship to be healthy. Therefore, in conjunction with the animal facility, the laboratory has decided to age female BALB/c mice instead.[1] Females of this strain are thought to be quite docile, unlikely to fight with one another, and so can be kept as a community within a cage for years.[2] Given that this is the first time this animal facility has aged female BALB/c mice, Janet says it has been a big learning process for them.[3]

Janet continues to explain to me that the animal facility has learned, in the process, that female BALB/c mice develop cancers with age, particularly ovarian and liver cancer. Janet and the other animal technicians began finding bloody discharge in the cage bedding of the mice. They became very concerned and called the veterinarians in, who started investigating the causes of death for these mice. It was at this point that the veterinarians realized that the mice were dying of cancer. Janet explains further that the veterinarians did not believe the mice were experiencing any pain or suffering from the cancerous tumors, and so the aging and immunity research could continue. The technicians, however, remained concerned. Janet explains to me that they began to check the cages of the aged, female BALB/c mice every day. She quickly adds that this form of care proved to be excessive. The BALB/c mice are fragile, and daily cage checks turned care into a form of surveillance that was too stressful for the mice. So now the techs check the cages very carefully once a week, and that is what I will watch Janet do today.

Janet starts by organizing the cages according to age, beginning with the oldest mice. All the cages with older BALB/c mice have a red tag attached to the front, stating: "Aging illness, bloody discharge." As we are looking at the lined-up cages, each with this red tag, Janet tells me that this is clearly an age-related disease. "Look," she says, "there's bloody discharge in every cage with mice over one year old."

* * *

I wrote these fieldnotes in the summer of 2016. I read them differently when I began writing this book in the autumn of 2020. I found myself writing about these mice and their cancers during the COVID-19 pandemic and the corresponding lockdowns in the UK during 2020 and 2021.[4] When I learned that a vaccine is the primary strategy for addressing this pandemic in the medium to long term, I could picture these geriatric mice and their role in pharmaceutical research. The COVID-19 coronavirus is far more devasting for older people. The fact that older people uptake vaccines less completely than younger people do became relevant to me in a new way in this context. These geriatric BALB/c mice were foundational to the production of a vaccine that would get us out of lockdown, even though experimental animals were conspicuously absent from the public discourse.[5] When I took my first vaccine for COVID-19 in 2021, as I continued writing this book, I reminded myself that I was ingesting the lives and deaths of those geriatric mice whose bodies most certainly—if circuitously—contributed to this vaccine. And yet the fact that these mice are part of this vaccine was very easy to forget, even for me, without putting some work into the act of remembering.[6]

This book is an act of remembering; my goal is to make often invisible laboratory animal bodies, as well as the care that goes into those animal bodies, a visible part of bioscience and biomedicine. I do this neither to reject using animals in biomedical science nor to critique vaccines. I have, after all, taken all of the vaccines offered to me by the National

Health Service (NHS) in the UK. Indeed, in order to study laboratory animals empirically as a sociologist, I could not be against animal research. Instead, I do this to posit that the very fact that we see laboratory animals as marginal to biomedicine is, actually, a rather surprising social accomplishment.[7] And this marginalization tells us something about not only biomedicine but also care and inequity.

I follow in the footsteps of feminist science and technology studies (STS) scholarship that describes invisibility as a social process, one that reproduces power relations and inequalities. Susan Leigh Star's collaborative research with Anselm Strauss (1999) on invisible work, as well as her own, separate research on invisibility (Star [1988]2015), highlights how certain types of work are publicly recognized, whereas other work is strategically hidden in a manner that tends to reproduce existing social inequities. Moreover, Adele Clarke's methodological and theoretical work on implicated actors (Clarke and Montini 1993; Clarke and Star 2008) draws attention to those implicated but not present in a situation of action. I am inspired by Monica Casper and Lisa Jean Moore's (2009) troubling of the dialectic of visibility and invisibility, showing how movement between being visible and invisible is interlinked with relative privilege and power. What are the consequences of Janet's care work, in which the geriatric mice that serve as surrogate humans in pharmaceutical research are an invisible part of bioscience and, thus, biomedicine? What do we learn about inequality when we make visible the work that is involved in making animals work for bioscience?[8]

It is important to ponder and question the scale of animal use that is part of biomedicine. In 2022, there were 2.76 million scientific procedures involving animals in the UK and 59% of those procedures used mice (Home Office 2023).[9] The scale of animal use in biomedical science matters for the animals (as animal rights activists draw attention to). This book addresses how animal use also matters for how the humanitarian aspects of biomedicine are practiced and understood. Whenever biomedicine becomes a salve for human suffering, that salve has relied upon the lives of other-than-human species. As such, humanitarianism does not simply apply the technical skills of medicine in treating people in emergency situations. Humanitarianism ingests the use of animals for the purposes of improvements in human health.

The core argument of this book is that a version of humanitarianism occurs in the everyday knowledge practices of biomedical science. I have given the name of "more-than-human humanitarianism" to this version in an attempt to better reflect its history and its present. Initially, the naming and description of a more-than-human humanitarianism is relevant for scholarly debates between animal studies and animal rights. More than this, it also expands how we think about doing relations with one another, human and other-than-human, in the context of entrenched inequalities—inequalities that humanitarianism is both a product of and response to. Animal research is a problem because it asserts an unjust inequity between human and nonhuman animal life. Many scientists, animal technicians, veterinarians, and animal rights activists believe we should replace animals in biomedical research—and I agree. These ideas are fairly easy to gain consensus upon. The more difficult question is: How? Empirical research is crucial to answer this question.

Understanding how people manage realities that are institutionally erased, in other words, rendered invisible, is therefore one of the goals of this book. To begin the work of clarifying the relationships between laboratory animals in medical research and humans in humanitarianism—as well as to name those connections as "more-than-human humanitarianism"—I draw on the long-standing links between "witnessing" in both humanitarian work and the ethnographic practices of social science. Witnessing is a practice associated with Médecins Sans Frontières (MSF), who distinguish their approach from the neutral and discreet approach of the International Committee of the Red Cross (ICRC). It is also one of the primary justifications for ethnographic methods, which this book uses (see Givoni 2016). Witnessing is a problematic practice, for it is neither neutral nor does it take a political stance vis-à-vis the polarized perspectives of being for or against the use of animals in research. This book posits that it is important to see the work involved in medical interventions that explicitly address human suffering but also to see the work involved in making those medical interventions possible—work that requires animals to suffer and die.

My research has been in close conversation with an interdisciplinary scholarship that has critiqued the "polarization cycle," wherein scientists and animal technicians are pitted against animal rights proponents such that objectivity and care are considered antithetical (G. Davies et al. 2024). In contrast to the dialectical model that the polarization model presumes, wherein two sides are pitted against one another, I have instead drawn on the "nexus" approach used in G. Davies et al. (2024)—a heuristic that assumes the networked, situational, and entangled metaphors that have been central to STS. I appreciate the critique that such models and positions risk legitimizing practices, through an emphasis on complexity, that other scholars critique so as to resist and abolish them (e.g., Giraud 2019, 2024; Peggs 2009, 2013). However, in this era of violent polarization, I think it may be worth remembering that polarization, too, can be otherwise.

A Nation of Animal Lovers?

Why, you may ask, ameliorate humanitarianism with a more-than-human clause? I am an empirical researcher, and the research that informs this book is rooted in the United Kingdom. Humanitarianism arose as an "in vivo code" or as an "actor's category." It proved necessary to understand humanitarianism in order to understand the use of laboratory animals in Britain.

This book comes out of over a decade of research on care for laboratory animals in Britain. It began as a pilot study, which provided the basis for a mixed-methods and multisited collaborative research project funded by the Wellcome Trust. Nathalie Nuyts, Juan Pablo Pardo-Guerra, and I worked together on a survey of British scientists and subsequent qualitative interviews.[10] Joanna Latimer and I worked "alongside" one another—conducting participant observation separately and simultaneously in different locations of the same Institute in a manner that was consistent with Latimer's (2013a) concept of "being alongside" that I use throughout this book.[11] Tarquin Holmes and I worked together on a historical analysis of the 1875 Royal Commission on Vivisection.[12] This book specifically draws upon the fieldnotes that I wrote as part of the ethnographic research I conducted as part of the pilot and the Wellcome Trust–funded research. However, the focus on humanitarianism resulted from the survey findings and qualitative interviews, shaping the historical research.

The survey, somewhat unexpectedly, found that being British was associated with reporting that animal care is a crucial part of conducting high-quality science. This was even though all the scientists who participated in the survey were working in Britain, and thus in the same regulatory milieu. Unsure how to interpret this finding, Nathalie Nuyts, Juan Pablo Pardo-Guerra, and I drew upon the concept of "civic epistemologies" (Jasanoff 2005; Prainsack 2006) and argued that "animals" and "care" in Britain may converge in taken-for-granted assumptions about what constitutes good scientific knowledge. We located this in the history of animals in humanitarian thought and action during 19th-century Victorian Britain. We suggested that care for animals in science is likely shaped by Victorian-era, class-based thinking, wherein the upper class has a duty to *protect* those who are vulnerable, including humans and animals (Friese, Nuyts, and Pardo-Guerra 2019 Online First). What we see here is thus a very specific notion of "care" as protection that is linked to hierarchy, wherein those who have the power to dominate must do so with responsibility. Keith Thomas (1983) has traced this ethos as far back as the medieval period, in religious doctrine particularly. This book explores the implications of this for how people relate with animals in laboratory science.

Rather than using ethnographic fieldnotes to describe the social space of a laboratory or an animal facility, I instead use ethnographic vignettes to describe affective moments that speak to what I am calling more-than-human humanitarianism. I interpret the significance of these interactional moments through amplification, using fiction to resonate more general concerns about inequalities, suffering, and injustice that these moments speak to. Juxtaposing ethnographic vignettes with fiction, I ask the following question: What does the laboratory animal as a relation look like in Britain through the "prism" (Svendsen et al. 2017) of humanitarianism, and conversely what does humanitarianism look like through the prism of the laboratory animal? I have built upon critical scholarship on humanism and humanitarianism to see fleeting moments of care and compassion that I experienced as part of the ethnographic research, moments that encapsulate the affects of more-than-human humanitarianism and that might otherwise go unnoticed.

I conclude that this book may be a kind of salvage sociology. I am not collecting indigenous culture or indigenous DNA from people deemed to be dying from modernization, as the problematic practice of salvage anthropology does (TallBear 2017). But it may be the case that I have collected the affects of more-than-human humanitarianism as a modern concept—one that has linked humans to the other animal species as much as it has divided humans and those species—just as this ethos is being hollowed out. My ability to name "more-than-human humanitarianism" may denote that this way of being in the world is disappearing, in Britain at least, or is changing into something related but also new.

Rethinking Humanitarianism

Humanitarianism is neither an innocent nor an inherently good concept for humans or other species.[13] Humanitarianism clearly reifies the idea of species difference, often resulting in a political orientation where we as humans are admonished to address the needs of our own species first before addressing the needs of other species. This idea has been rightly, in my view, critiqued across human rights and animal studies. Not only does human exceptionalism reify a hierarchy between humans and nonhumans, but it also works to legitimize hierarchies between humans on the basis of race, sex, age, and ability. I take it for granted that the "human" of humanitarianism, human rights, and humanism does not guarantee freedom from violence or discrimination within the species because the very concept itself is a hierarchy, one that allows for violence and discrimination as a dividing strategy (Cubukcu 2017; Wolfe 2010). Putting the more-than-human clause before humanitarianism does nothing to change this history of noninnocence. Yet, by recognizing the limits of humanitarianism for humans and other animal species alike, I seek to show important benefits and strengths of this concept as well.

When we think about humanitarianism, we tend to think about medical treatment, food supplies, temporary housing, and sanitation for people in an emergency situation—one that has been caused by war, violence, a "natural" disaster, or some combination of those. However, humanitarianism also developed and changed dramatically across the 20th century in ways that challenged this focus on reacting to emergencies. Alleviating suffering has increasingly been considered insufficient for many involved in humanitarian aid, and emphasis has been placed on addressing the structural elements that give rise to any disaster. The lines between humanitarianism and human rights have therefore been blurred (Rieff 2002). If humanitarianism is about saving lives in a crisis situation as an outgrowth of charity, human rights has a longer time horizon and a political orientation based on addressing the root causes of suffering (Krause 2014, 149). However, this politicization of humanitarianism, both within relief organizations and by nation-states and other political actors, has challenged the project of humanitarianism and has created fissures within the field. David Reiff (2002) has, for example, argued that Kosovo marked the death of humanitarianism through the development of "humanitarian wars" that continued into Iraq and Afghanistan and persist today.

What has distinguished humanitarianism from welfare is its orientation beyond the nation-state—care and compassion for a *distant* other who suffers is seen as central to humanitarianism (Barnett 2011; Reiff 2002). In comparison, animal welfare, which became increasingly institutionalized in the mid-20th century, put laboratory animals within the nation-state (Svendsen 2022; Kirk 2014). However, as Mette Svendsen (2022) points out, being near-human means that animals are always also at some distance. Therefore, the *distant* other can be thought of not only spatially as in humanitarianism, but also biologically, informing how animals became entangled in the history of humanitarianism in Britain in the 19th century. This focus on distance as both spatial and biological shaped the ways in which animals were included in humanitarianism in Britain during the 19th century, and so I turn to that history now.

Across the 19th century, several laws reformed how animals were treated in Britain, starting with cattle in 1822 and culminating with research animals in 1876 (French 1975). During this time, charges of cruelty toward animals were largely directed at Britain's lower classes and colonized subjects; learning to care for animals was thus seen as a civilizing process (Tague 2015). Laboratory animals were, therefore, a particularly troubling intraclass concern (French 1975). Protecting vulnerable others through benevolent paternalism became a priority in Victorian-era Britain, one that linked humanist concerns about animal welfare with child welfare and the abolition of slavery. The upper classes asserted their moral superiority through their care for their charges—human and nonhuman alike (Ritvo 1987). Animals generally, and laboratory animals specifically, were thus actors in fomenting humanitarian thought and action in Britain during the 19th century.[14]

The humanitarianism that most Britons are familiar with today became explicitly focused on human suffering, as separate from animal suffering, only in the context of the First World War.[15] The human subject of humanitarianism was detached from the animal, as the brutalities committed among humans pushed the animal out of sight (Barnett 2011). However, those concerned with laboratory animals in Britain remained attached to humanitarianism for much of the 20th century. For example, Russell and Burch consistently used the term "humanitarian" in their *Principles of Humane Experimental Technique* ([1959]1967) that first introduced the concept of 3Rs—replace, reduce, and refine animals from scientific research—which is the current-day gold standard in doing animal experimentation in an ethical manner worldwide (G. Davies et al. 2018; Kirk 2018). And where 19th-century antivivisectionists were part of other humanitarian campaigns (French 1975), animal activism has worked since the 1970s by aligning itself with human rights (e.g., Singer [1975]1995).

Since the 1970s, animal rights activism has sought to create a corollary to human rights through a politics of inclusion wherein nonhuman species with traits or behaviors seen as commensurable with humans (e.g., primates) have been argued to have legal rights and protections (for discussion, see, for example, Wolfe 2003a, 2003b; Cochrane 2012; Acampora 2006). However, this is where the stories of humanitarianism and more-than-human humanitarianism diverge. If humanitarianism and human rights aligned and blurred across the 20th century (Rieff 2002), animal rights has been in a more antagonistic relationship with veterinarians and animal technicians who practice animal welfare. In other words, animal rights and animal welfare are in tension with one another, in what Eva Haifa Giraud (2024) calls incommensurable care. Animal rights has emphasized the abolition of animal use, critiquing the ways in which care for laboratory animals perpetuates this practice.[16] More-than-human humanitarianism, however, is rooted in the crises that arise when one is in the presence of animal use. Why I refer to this as more-than-human humanitarianism, rather than animal welfare, is to call out and name the ethos that surrounds animal welfare and that goes beyond a scientific rationality in its practice. This allows me to emphasize the historical genealogy of these affects in 19th-century British humanitarianism that included animals quite explicitly.

More-than-Human Humanitarianism

In this book, I focus on what it feels like to care for laboratory animals when that togetherness is stage-managed so that the animal lives and dies in a very specific way.[17] In the process, I consider everyday practices as being in tension with ideological concerns. I therefore engage in a set of debates that reverberate across animal studies and social studies of humanitarianism and human rights. How do we attend to everyday practices of care and appreciate these practices, while also holding on to the fact that these are sites where hierarchies are also socially reproduced?

The idea of "partial connections" (Strathern 2004, 1997, [1992]1995) plays a central role in how I analyze the relationship between humans and animals in humanitarianism. In contrast to the notion of hybridity, where two different kinds of things are mixed, Marilyn Strathern ([1992]1995) has argued that English thought is merographic, meaning that two different kinds of things connect in a concept while remaining different. Where hybridization focuses on mixing and dialectics focuses on a synthesis, merography focuses on connection across difference. Kinship is Strathern's (2020, [1992]1995) key example of merographic thinking, and it is useful to demonstrate the point. Kinship is a concept rooted in British anthropology and English family life, and it is a concept that connects biology (i.e., born) with society (i.e., bred). However, biology and society are not collapsed into one another with kinship; they are related and the form of the relation across these different spheres is what is important. I understand humanitarianism as another site of merographic thought in Britain in that it connects the human and animal without collapsing the human and the animal into one another.[18]

This approach has also been deeply informed by Joanna Latimer's (2013a) concept of "being alongside." Latimer articulates the problem of the human–animal divide as a historically produced concept and practice while also providing a means to think outside of that problem in a manner that explicitly draws upon Strathern's (2004) "partial connections." Latimer develops "being alongside" as an alternative to Donna Haraway's (2008) "becoming with."[19] She distinguishes "being alongside" for its conceptual basis in "partial connections" (Strathern 2004), where Haraway's "becoming with" is based on hybridity. While both Haraway and Latimer draw on the work of Strathern in developing their arguments, Latimer (2013a, 80) focuses specifically on how Strathern emphasizes that relations involve not only attachment and connection but also detachment and disconnection. More-than-human humanitarianism, as it is practiced in British laboratory science, is rooted in this simultaneity of connection and disconnection.[20] Being alongside allows us to see how humans and animals are related and relating to one another, in practice and ideology, without collapsing key differences in the process.

By exploring animal welfare as an instantiation of more-than-human humanitarianism—one that has developed in relationship to human humanitarianism and in contrast with animal rights—my goal is to present a set of practices and ideas that have otherwise been invisible. My goal is descriptive rather than normative. More-than-human humanitarianism is, therefore, *not* a utopian concept, in the way that, for example, Donna Haraway's (2008) companion-species concept is. More-than-human humanitarianism is my way of naming a set of ideas and practices I have witnessed that speak to ongoing concerns within both social studies of humanitarianism and human rights as well as animal studies and animal rights. This book considers both the benefits and the limitations of more-than-human humanitarianism for those who care about inequities and suffering both within and between species—and where those cares intersect and where they don't.

Laboratory Animals

This book focuses on those animals that serve as surrogates or substitutes (C. Thompson 2013; Svendsen 2022) for humans in biomedical research and the people who both care for and kill these animals. Within the biomedical domain, laboratory animals are constituted as sacrificeable species that can stand for and represent human bodies and diseases (Lynch 1989; Birke, Arluke, and Michael 2007; Svendsen and Koch 2013). Laboratory animals have thus been deemed "killable" (Haraway 2008). The use of animals as experimental models in biomedical research is justified by this distinction between species (Ingold 2011)—nonhuman animals can be modified, experimented upon, and eventually killed in ways that humans can and should not (Svendsen and Koch 2013).[21] At the same time, however, animal modeling also troubles the idea of a clear species delineation because of a presumed interspecies homology in molecular substance (genomes), process (such as reproduction, development, and aging), or both that makes modeling relevant (Svendsen and Koch 2013). In this context, the differences between humans, experimental animals (e.g., sentient species like mice, pigs, cats, and dogs), as well as other living experimental organisms assumed to be insentient (e.g., flies, worms, yeast, bacteria) are generally emphasized, while always also open to question through comparison (Ankeny 2007; Friese and Latimer 2019; Friese and Clarke 2012).

As animals have served as surrogate humans, the question of pain and suffering has been central (Shmuely 2023). Indeed, it was a concern about pain and suffering that motivated many humanitarians in the 19th century to advocate against vivisection in Britain. Tarquin Holmes and I (2020) have explored this, based on Tarquin's analysis of the 1875 Royal Commission on Vivisection. This commission led to the Cruelty to Animals Act of 1876, which marked a key turning point in the use of animals for laboratory science. It was the first legislation to regulate the use of animals for science. It established a stringent licensing system still used today to govern how scientists use animals. Holmes and I have argued that the anesthetized animal represented a "boundary object" (Star and Griesemer 1989) through which stakeholders articulated and contested the morality of animal life and death in science. We traced how anesthetics came to be viewed as a prophylactic against unacceptable abuses in the laboratory, addressing animal suffering while allowing for the continued practice of physiological experimentation. Ensuring that animals in science are not suffering—or are not suffering excessively—has become the primary legitimization strategy that allows the use of animals to persist in science. It is not uncommon for people to contrast the relative suffering of a prey species, such as mice, in laboratories as significantly less than what prey species, such as mice, experience in other milieus, such as our homes.

For much of the 20th century, the focus thus came to be on ensuring that laboratory animals were sufficiently protected from scientists. Social science research supported this need for this protection, as researchers found that scientists learn to distance themselves from laboratory animals as part of their education so that they understand animals as "tools" rather than sentient creatures (Birke, Arluke, and Michael 2007, 11, 14). Animal care work was professionalized in this context (Kirk 2010, 2014, 2008, 2012; Druglitrø 2018), but as a service to science and was thus relatively marginalized. This marginalization was evidenced by the systematic erasure of animal husbandry practices from scientific journal articles (Lynch 1989; Holmberg 2011; Birke, Arluke, and Michael 2007; Lederer 1992). In this context, animal husbandry (i.e., the work involved in feeding, housing, handling, and reproducing laboratory animals) has been thought of as an extrascientific concern, not part of the work that scientists do themselves but rather managed by animal technicians and veterinarians (Holmberg 2011; Birke, Arluke, and Michael 2007; Greenhough and Roe 2011). Indeed, previous research has indicated that scientists do not see animal care as part of science (Lynch 1989), and notions of objectivity have been used to support this (Birke, Arluke, and Michael 2007). These workplace hierarchies help explain how and why the role of laboratory animals, and the work of those who care for laboratory animals, has come to be an invisible element of the "medical–industrial complex" (Estes, Harrington, and Pellow 2001) that focuses on universal knowledge and standardized therapeutics (Friese, 2024).

At the turn of the 21st century, however, it was becoming increasingly common to hear life scientists in Britain say that scientific research relies upon high-quality laboratory animal care (G. Davies 2010; Friese 2013; Hurst and West 2010). The idea here is that "happy animals" make "good science" by introducing fewer confounding variables into research (Poole 1997). The introduction of the widely cited "Animal Research: Reporting of *In Vivo* Experiments" (ARRIVE) guidelines, which require authors to report animal husbandry practices in their scientific journal articles, attests to this increasing focus on animal care within science at the institutional level (Kilkenny et al. 2010). There is a growing discourse positing the need to create a "culture of care" in laboratories and animal facilities in order to ensure the well-being of animals used in research. Here change is sought at the organizational level so as to exceed animal welfare (e.g., regulatory) requirements (Klein and Bayne 2007; G. Davies et al. 2018). The research informing this book started in 2009 and has sought to track these developments.

Materials and Methods

This book draws specifically upon the fieldnotes I wrote as part of the participant observation research I conducted during the pilot project and

Wellcome Trust–funded research.[22] I analyze these fieldnotes by drawing upon and adapting Mette Svendsen's (2022) and her colleagues' (2022) prism ethnography method. They use prism ethnography to juxtapose related but different ethnographic sites (e.g., a laboratory using pigs as a model for preterm infants in the ICU) by asking what site A (e.g., the laboratory) looks like through the prism of site B (e.g., the ICU) and vice versa. I use Svendsen's prism methodology to ask what laboratory animals look like through the prism of humanitarianism and what humanitarianism looks like through the prism of laboratory animals. Where Svendsen takes a spatial approach rooted in ethnography, I have instead focused on situations (Clarke 2005) or fields (Bourdieu 1987) of meaning and practice as my unit of analysis. In other words, I have sought to understand the situation of using animals in the field of biomedical science rather than seeking to understand the social space of the laboratory or the animal facility.

Prism ethnography works through juxtaposition rather than comparison. The goal is not to compare sites A and B by asking how they differ, but to ask what difference is made by bringing these sites together and reading them alongside one another. Svendsen's prism ethnography inherits Marilyn Strathern's (2004) concept of partial connections, in that the human and the animal are connected in biomedical science and practice without collapsing into one another. Partial connections rather than hybridity is the metaphor that juxtaposition works through.

Prism ethnography, to my mind, is thus also related to Karen Barad's (2007) method of diffractive reading through this practice of juxtaposition. Diffraction similarly asks what difference is made by juxtaposing different kinds of texts. Kalindi Vora (2015) developed this method for ethnographers specifically, as she juxtaposed empirical research with fictional accounts to empirically capture the productivities of fantasy and the fictive for doing analytic work. Like Vora, I seek to amplify moments of care and compassion by juxtaposing ethnographic fieldnotes with selected fictional texts. I explore specifically *Transcendent Kingdom* by Yaa Gyasi, *Disgrace* by J. M. Coetzee, and *Washington Black* by Esi Edugyan. Just as the scientists I observed used recombinant DNA technology to amplify the DNA in a gene of interest so as to study it, I use fiction to amplify the significance of interactions so as to study them. I thus develop amplification methodologically, and speak to the growing use of fiction to do social theorizing (Vora 2015). Using fiction allows me to interpret and render meaningful what might otherwise be considered fleeting, interactional moments. Fiction has helped me to express, in language, moments of profound empathy, compassion, and understanding that I experienced while conducting this research.

There is a moral rationale for juxtaposing ethnographic fieldnotes with literary fiction as well. If animal rights are built upon law to instantiate formal ethics, this book builds upon fiction to instantiate a virtue ethic that draws upon Martha Nussbaum's *Poetic Justice* (1995). Nussbaum has argued that fiction, in which people step outside of themselves and imagine what someone else's life feels like, is an important place where compassion is taught. Nussbaum's work is foundational for an argument this book makes—that both formal and virtue ethics are required and practiced with laboratory animals. Virtue ethics, however, risks being eroded without recognizing the time necessary for both its practice and its acknowledgment.

I have chosen novels that I believe address a problematic component of animal studies, a field that I work in and contribute to. Animal studies have been dominated by white scholars, and white women in particular—a demographic that I fit into. The problem here is the history of race, racialization, and racism that animals have been embroiled in, through the entwining of the animal and animalization with the creation of race and racialization in the twin processes of colonization and capitalization (Lundblad 2013). In 19th-century Britain, evolutionary thought linked the human with the animal through the idea of race to connect some humans more closely with animals in ways that justified hierarchies of oppression and subjugation through dehumanization (Haraway 1989). To make this history present, I have chosen three novels that make race, gender, science, and animals central to their storytelling. In the process, I attempt to address social structures and oppressive histories without reducing the complexity of any person to those histories.[23]

In selecting novels, I have been inspired by Joshua Bennett's (2020) scholarship, in which he problematizes the links between racialization and animalization that took shape in the context of 19th-century European imperialism in tandem with natural history and evolutionary theory. Bennett (2020, 4) has argued that there is a reconstructive turn within black literary studies toward novels that both critique Western philosophical thought entwining animalization and racialization *and* articulate interspecies relations in new ways that "abolish the forms of antiblack thought that have maintained the fissure between human and animal." I have selected *Transcendent Kingdom* and *Washington Black* as novels I believe fall into Bennett's category of this reconstructive turn.

Bennett describes his methodology as one that asks (2020, 5): "How have black authors cultivated a poetics of persistence and interspecies empathy, a literary tradition in which nonhuman—and thus also, ostensibly, non-thinking—life forms are acting up and out in ways we might not expect or yet have a language for?" I draw on Bennett's methodology to see relations and ways of being with other species in the everyday practices of science that might otherwise risk being left unseen and thus invisible. I do not seek to valorize these moments as somehow utopian but rather to see them as containing kernels of possibility for understanding interspecies relations differently if they are amplified. Concurrently, I keep in mind and address the structural limitations of these encounters as necessary to this method (G. Davies et al., 2024). In other words, I have sought to find ways to see ethnographic moments as both embedded within histories of subjugation while also attempting to understand "the laboratory animal" in a way that differs from, or stands alongside, the use value of animals. I contend that we might be able to see the practices of caring for laboratory animals through other lenses and discourses, and these lenses and discourses may be beneficially amplified. Through this I hope to show the benefits and limitations of what I call more-than-human humanitarianism.

In becoming entangled with life scientists and animal technicians, my research will always foreground the complexity of animal use and will therein fail to condemn it. Being present in the laboratory means that scholarship such as mine, as Eva Haifa Giraud (2019, 2024) has shown, is necessarily *not* entangled with animal rights activism. For Giraud, her politics means that she cannot and will not become entangled in animal use because her resistance requires retaining this distance. For a researcher, becoming entangled in one set of worlds means exclusion from others. Thus, there is tension between animal studies that seeks to understand how scientists and technicians care for the animals they kill and animal studies that aims to abolish that science for its hierarchizing of human interests. To address these debates, I also look at the novel *Disgrace* specifically as it has been contested along these lines.

Structure of the Argument

The chapters are organized according to key discursive anchors for both humanitarianism and laboratory animal welfare, which include: suffering, care, killing, sacrifice, compassion, and consent. Each chapter is rooted in ethnographic vignettes that seek to convey the affective qualities of everyday enactments of these discourses, which might otherwise be ephemeral and even prelinguistic. These ethnographic vignettes are given meaning through amplification, wherein the novel is discussed in juxtaposition to speak back to that theme.

One novel is used across two different chapters that address distinct but related themes. Suffering and care are, for example, particular but also very much connected, and I analyze one novel across these two chapters to mark out these connections. Yaa Gyasi's novel *Transcendent Kingdom* is analyzed across chapter 1 on suffering and chapter 2 on care. Esi Edugyan's novel *Washington Black* is analyzed across chapter 5 on compassion and chapter 6 on consent. Both of these novels were chosen as instances of what Bennett (2020, 4) has framed as both critiques of Western Man (Wynter and McKittrick 2015) while also articulating interspecies relations in new ways. Through juxtaposing these novels with ethnographic material I show how interspecies relations are conducted in science in ways that certainly reproduce the hierarchical use of animals for human betterment while also foregrounding how relations in laboratory science also exceed these instrumental and utilitarian framings. Meanwhile, *Disgrace* by J. M. Coetzee is used to amplify the theme of killing in chapter 3 and sacrifice in chapter 4 while also intervening in a set of debates about the ethics and politics of animal studies.

Each chapter concludes by considering the implications of the more-than-human approach to humanitarianism, as detailed across the chapter, for humanitarianism writ large, which is human-centric. I emphasize in chapter 1 that suffering is something that needs to be recognized and gone through, and so "alleviating suffering" as a trope for humanitarianism could create certain blind spots as well as sites of denial. In chapter 2, I emphasize that care is a knowledge practice that requires a togetherness that lasts, and that raises questions about how humanitarianism moves from place to place and from crisis to crisis. Chapter 3 ponders the social fact that some animals and some humans have been deemed killable, under certain conditions and by certain authorities. In this context, I raise the very uncomfortable social fact that "killing," or bringing to an end, unconscionable social relations (e.g., the laboratory animal as a human–animal relationship; apartheid in South Africa) has often coincided with the unconscionable killing of individuals. Chapter 4 concludes with the social fact that both animals and humans are routinely called upon to sacrifice themselves. It is the point that Derrida himself ended at: sacrifice should not exist, and yet there seems to be no way out. Chapter 5 argues that where humanitarianism often starts with compassion, we could productively see it instead as a privileged *outcome* of togetherness. Chapter 6 asks if the receiver of humanitarian aid can say no to the aid provider and if the giver of aid is equipped to be able to hear a "no" that may not be spoken in language.

I conclude the book by returning to the question of what naming more-than-human humanitarianism does for practices in caring for humans and animals. What are the benefits and limitations of this ethos? What are the implications of these benefits and limitations for humanitarianism itself? In answering these questions, I return to the methodological nationalism (Wimmer and Schiller 2002) that informs this book, and consider what this research on Britishness means for debates over the politics of national identity formation, social reproduction, and social change. To address the theme of social reproduction and change, I ask what naming "more-than-human humanitarianism" denotes in the context of the partial-connections method of this book. Strathern argues that merographic thinking works to make the implicit explicit, and that this has been the goal of British academic thought, at least British anthropological thought. But the process of making the implicit explicit means that the very concept under investigation may actually be slipping away. Through Strathern, I conclude by asking: What does it mean to make more-than-human humanitarianism explicit in Britain today? As I write, the Tory government is working hard to remove itself from the European Convention on Human Rights in order to remove refugees to Rwanda in a broader national politics that is rooted in the logics of divide and rule at home and abroad. In this context, I consider what we might want to salvage from more-than-human humanitarianism, which I contend offers a means to enter the thick of ethics in practice, without necessarily providing a way out. To my mind, this is the key benefit of more-than-human humanitarianism, and its key limitation.

1

Suffering

Shortly after moving to the United Kingdom from the United States in 2009, I gave a seminar on the ways in which care work is marginalized in life science research conducted in zoos—to the detriment of science itself. This talk was based on my sociological research on cloning in zoos, and I thought this finding was institutionally unique because "wild" animals are not "model organisms."[1] A biomedical scientist who works in a British university approached me after my talk. Elspeth, which is the pseudonym I have given her, told me that she believed the marginalization of animal husbandry and care was not unique to zoos but rather was a barrier in life science research more generally.[2] She also thought the marginalization of care work was creating a barrier to translational medicine. Suffering animals result in stressed animal bodies, and this ultimately creates a series of confounding variables in any physiological research.[3] To further explain this, Elspeth invited me to her laboratory so that I could see how she was incorporating improved animal care into the experimental system she had created to study and produce drugs to treat cardiovascular dysfunction. This visit was my first introduction to laboratory animals, and I here came to see that responding to a suffering animal with care is a hard-fought goal that needs to be legitimized and sustained time and time again.

 This chapter explores how I was shown laboratory animal suffering and responses to it in the contemporary moment, which occurs at different scales (e.g., individual and population), temporalities (past, present, and future), and embodiments (e.g., haptic and distanced). I explore this through the initial pilot research I conducted in a university laboratory as well as in my later ethnographic research at the Institute. I amplify the significance of these experiences by juxtaposing my ethnographic storytelling with Yaa Gyasi's fictive storytelling in the novel *Transcendent Kingdom* (2020). Just as Elspeth brought me into her lab, prompting my interest in these themes, Gyasi's novel was inspired by her best friend's laboratory research. Fiction helps me to articulate and value structures of feeling that I experienced as circulating in laboratories, but that are often difficult to bring to the surface. With the help of Gyasi's novel, the humanitarian aspects of doing a science that *both* responds to *and* creates suffering—that I witnessed in my participant observation research—becomes affectively articulable.

I situate this analysis of suffering in a history of British humanitarianism that has responded to the physical suffering of both human and nonhuman animals. Nineteenth-century humanitarianism in Britain was responding to and acting against gratuitous and extreme forms of physical suffering rampant among humans and nonhuman animals alike. I use fiction in my analysis here in part because fiction also played a role in the work of making suffering visible so as to ameliorate it in the Victorian era. The living and working conditions of the poor in Britain were the topic of concern for Charles Dickens in his novels, in which he sought to represent suffering in a bid to change its material production. Harriet Beecher Stowe's *Uncle Tom's Cabin* ([1852]1999) made the horrors of slavery visible as part of the work toward abolition. Anna Sewell's *Black Beauty* ([1877]2014) represented the kinds of vulnerability and suffering that working animals experience at the hands of humans, as legislation was being made in Britain to govern, in tandem, the treatment of working-class people and their animals. Early humanitarians often worked across different sites of suffering, resisting slavery, child labor, and the vulnerabilities of the poor, women, mentally ill and animals. Jeremy Bentham's (1789) famous statement that continues to inspire animal rights and animal welfare alike was in part inspired by the abolition of slavery in France: "The question is not, Can they reason?, nor Can they talk? but, Can they suffer? Why should the law refuse its protection to any sensitive being?"

Suffering Animals and Humanitarianism, c. 1875–1986

As part of the research that this book has developed out of, the historian and philosopher of science Tarquin Holmes conducted historical research on laboratory animals in 19th-century Britain. I have learned much from our writing together, and I summarize some of this historical research (Holmes and Friese 2020) to situate the question of animal suffering today.

Vivisection is the term used to describe experiments conducted on a living animal, and up until the 19th century this was generally done without anesthesia. William Harvey used vivisection to study the circulation of blood in the 17th century. Robert Hooke used animals to study respiration in the same period. Anita Guerrini (2003, 400–402) describes how, in one of Hooke's experiments, a dog was respirated by bellows with its chest opened to expose its heart so that the irregular heartbeat could be seen when the bellows stopped. Guerrini emphasizes that Hooke was concerned about the suffering he was inflicting, citing him as having written in a letter to Robert Boyle: "I shall hardly be induced to make any further trials of this kind, because of the torture of the creature . . . the enquiry would be very noble, if we could find any way so to stupify the creature, as that it might not be sensible, which I fear there is hardly any opiate will perform" (Guerrini 2003, 401). But the experiments continued, although their extent did not increase significantly until physiology expanded.

The expansion and professionalization of physiology coincided with the expansion of vivisection in Europe and then Britain. In Britain, this coincided with a growing humanitarian critique of animal mistreatment. Richard French's (1975) history of antivivisection in Britain tracks in detail these developments across the 19th century.[4] He shows that antivivisection was an explicitly humanitarian concern motivated by religious arguments (see also Thomas 1983), as were humanitarian sentiments more generally in Britain. Across the 19th century, a string of legislative reforms served to protect animals as part of this humanitarian ethos, starting with slaughterhouses and cattle markets and expanding to bull-baiting to eventually include all livestock and then all domesticated animals. Laboratory animals were the final human–animal relationship regulated in Britain with the 1876 Cruelty to Animals Act.

The rise of vivisection and antivivisection in 19th-century Britain coincided with an increasing sense of health as a right among the British public, according to French (1975). Anesthesia, vaccines, and antiseptics were being put into use by physicians who were increasingly professionalizing. French (1975) notes that while much of the British public wanted these therapeutics even if they required vivisection, antivivisectionists held a different attitude toward health. French states (1975, 306): "Disease was the divinely ordained consequence of sin and folly, and was to be borne as such. . . . Experimental medicine based upon the plea of practical utility . . . meant that man was unwilling to bear his allotted share of pain in the world." Antivivisectionists sought a focus on prevention. Given the religious basis of the movement, this included not only sanitary measures but also "moral improvement." For antivivisectionists, human suffering should not be alleviated through animal suffering but rather through a change in human behaviors both structurally and individually (see Holmes and Friese 2020).

The nationalist stereotype of Britons as having a particularly strong love for animals gained solidity over the 19th century. Harriet Ritvo notes that, at the beginning of the century, the English

> "would have been surprised to hear themselves praised for special kindness to animals. They were surrounded by evidence to the contrary but that by the end of the nineteenth century a humanitarian crusader proclaimed a "sentiment of tenderness for those of the sentient lower creatures . . . has become an element in the spiritual life so strong that the continual violation of social obligations to them is a cause of pain and revolt." (Ritvo 1987, 125–26)

In her study of pet keeping in 18th-century England, Ingrid H. Tague (2015) similarly finds that pet keeping was viewed as a luxury at best, and even a sin, at the start of the century but had become a sign of moral virtue by the end (see also Thomas 1983). The idea of a particularly British love for animals is entangled with British industrialization, colonialism, and imperialism.

The animal legislations instituted in Britain across the 19th century consistently equated animals, through a hierarchizing logic, with the working class and racialized, colonized people. This is a process that is today known and critiqued as one of animalization. Yet the idea that animals had to be protected from the working class and colonized was also promoted; learning to care for animals was a "civilizing process" (French 1975; Ritvo 1987). Love for animals grew during the same period and at a similar pace as the "revolution in moral sentiments and the emergence of a culture of compassion" (Barnett 2011, 49) in Europe more generally. The purported superiority of humans to animals, an idea made possible by Christianity and the natural sciences, made alleviating animal suffering a moral responsibility of those who dominated. This paternalistic care for those who are more vulnerable paralleled while also reinforced a discourse of duty in the configuration of classism in Britain and racism in the British colonies.

Therefore, using laboratory animals in science fostered a particular kind of tension in Britain's class and racial politics. Britain developed physiological knowledge later than Germany and France, so these tensions were minimized for much of the 19th century. However, the tensions came to the fore as physiology gained a place in British universities and as science was rapidly professionalizing. The first chair for physiology was established in 1874, the year before the Royal Commission was charged with studying whether and under what conditions animals could be experimented upon. Holmes and I have argued that the anesthetized animal became a "boundary object" (Star and Griesemer 1989) in the context, allowing for physiology to persist in using animals for research without inflicting pain (Holmes and Friese 2020).[5] Where anesthesia allowed for living animals to be experimented upon without inflicting suffering, death was used to ensure that suffering would not persist after the experiment. The 1876 Cruelty to Animals Act thus required that laboratory animals be killed to prevent a recovery from surgery that would mean living a life in pain. Dmitry Myelnikov (2024, 37) states regarding the act that "death was a preferable outcome to pain and suffering in its logic." The act ended the reuse of animals in order to delimit a single animal's suffering in this context. Alongside anesthesia, humane killing became and remains the primary practice that veterinarians use to address animal pain and suffering not only in laboratory animal science but also in small-animal veterinary practices (Friese 2016).

When the 1876 Cruelty to Animals Act was updated with the Animals (Scientific Procedures) Act of 1986 (ASPA), pain remained a contested issue. Myelnikov (2024) has analyzed the politics of consensus embodied by ASPA, showing how this entailed a political puzzle given that Margaret Thatcher was prime minister at this time. She instantiated a fierce contempt for this long-standing approach to British politics (see also Kirk 2018). Myelnikov shows that the government's Committee for Reform of Animal Experimentation (CRAE)—in response to the resurgence of animal rights activism of the 1970s alongside several public controversies about animal welfare—engaged with moderate animal welfare groups and scientists. CRAE also excluded what it saw as more radical animal rights activists. An alliance between CRAE, the British Veterinary Association (BVA), and the Fund for the Replacement of Animals in Medical Experiments (FRAME) played a key role (Myelnikov 2024, 36; see also Friese and Nuyts 2018). The veterinarian became the key figure with experience in seeing, assessing, and interpreting animal pain and, therefore, ameliorating animal suffering. A "fragile" consensus existed between science and animal welfare while excluding animal rights (Myelnikov 2024).

ASPA, which regulates the use of laboratory animals in the United Kingdom, was updated in 2013 to align with the European Union Directive 2010/63/EU legislation on the protection of animals used for scientific purposes (Dennison and Petrie 2020). The regulations in the UK and the EU therefore align, and are considered the strictest for the use of laboratory animals in the world. These regulations that emphasize centralized control differ from those in the United States, where local flexibility is prioritized (G. Davies 2021). Notably, the United States Animal Welfare Act excludes rats, mice, and birds (Birke 2003) and represents a stark difference between UK/EU and US regulations. Lesley Sharp (2019), for example, recounts how US scientists receiving animal welfare training are presented with a PowerPoint slide in which a picture of a mouse contains the caption: "This is not an animal." That said, all US publicly funded research using mice, rats, and birds must be approved by the local (rather than centralized) Institutional Animal Care and Use Committee (IACUC). Gail Davies (2021) shows that centralization of regulations in the UK and EU brings in more stakeholders in setting animal welfare standards, while the US relies more heavily upon veterinary expertise. Using Singapore as an example, Davies further argues that other countries are not simply adopting either the US or the UK/EU regulatory approach, but rather adopt parts of both regulatory systems in line with unique local, civic epistemologies. The regulatory milieu in which the mice and rats in this book are discussed is therefore very different from the regulatory milieu in which mice and rats would be positioned in the United States or in other parts of the world. The relevant regulatory milieu discussed in this book is similar to those in Europe. However, the affects of more-than-human humanitarianism described in this book may not arise in Europe in the same kinds of ways.[6]

With this history in mind, I now turn to my first visit to Elspeth's laboratory in 2009, following her invitation to show me how care is marginalized in science.

Vignette

Upon my first visit to Elspeth's laboratory, I was shown the experimental system she had spent the early part of her career creating, which helped her establish a position at a leading British university. To help understand the significance of this new experimental system, I was also shown the traditional system she was trying to replace. Through comparison, Elspeth demonstrated that marginalizing or emphasizing animal care changes the setup of an experimental system.

Elspeth began by showing me her experimental system. Here, five plastic boxes, each containing one mouse, were lined up on a raised shelf in a semidarkened room. Underneath each box was a black metal plate. A computer sat on the desktop below. The mice were running around their cages, shredding and burrowing paper lining or playing within cardboard tubing. Elspeth explained that each mouse had a telemetry device implanted within its body. Telemetry is a technology that allows data to be wirelessly transmitted across distances. Given how much the mice were moving around, it was at first hard for me to see the device. But after a while, a mouse began to climb up the side of the cage, exposing its tummy in the process. "There," Elspeth said, "you can see it there." I saw a bump about a half-inch long protruding from the mouse's underside. "That's it," she told me. Elspeth went on to explain that her experimental system requires surgery under anesthesia, and so each animal is subsequently given analgesia to reduce pain. Nonetheless, Elspeth stated that the mice in her experimental system did not show signs of infection and behaved according to expected daily patterns.

Elspeth then led me to another laboratory containing three separate but interrelated rooms, one of which held rats used to study sepsis, which damages the cardiovascular system, with the traditional experimental system. Here, five somewhat larger plastic boxes were similarly lined up, each containing one rat.[7] Each rat was lying on its stomach in the middle of the box, with a somewhat flexible metal tube attached to its back. The tube was connected to a black box on a shelf overhead, and each black box was connected to a computer at the end of the room. Large black stitches held the tubing in place on the rat's back. The skin around the tube was very red, and I asked if this was infected; Elspeth replied that it could be.

Elspeth proceeded to demonstrate these animals as suffering. She put her hand in one cage, trying to call the rat to her with a "tsch-tsch." The rat did not respond. The same thing happened when she tried to call another rat to her. After multiple failures, Elspeth commented that rats should not act like this; they are friendly animals and should come to you when called. Indeed, I know that Elspeth enjoys working with rats for this very reason: because they are inquisitive in their interactions with humans. Based on this abnormal behavior, Elspeth contends that this experimental system is inappropriate from both a welfare and a scientific perspective. To her mind, the physiological consequences of stress and suffering have not been appropriately accounted for in this experimental system. As such, Elspeth wonders about the quality of the findings produced with these animal bodies.

Elspeth's uptake of telemetry cannot simply be explained by the availability of new and better technology. She was motivated to replace the tether because she did not want to work—day in and day out as a postdoctoral researcher—with animals subjected to the tether, which she believed caused physical and emotional suffering to the mice and rats she was working with. Locating her "shared suffering" (Haraway 2008, 82) with the mice and rats in the tether, Elspeth sought an alternative means to set up her experimental system. She understands this move as being motivated by a concern with both animal welfare issues and knowledge production. In terms of the latter, Elspeth found it difficult to differentiate the physiological response to stress and pain caused by the tether from the physiological response to the intervention that was being tested.

I have since seen Elspeth show videos of these two experimental systems at conferences to demonstrate the utility of telemetry as a technology of care that produces better scientific findings. She shows an image of playful mice with an implanted telemetry device contrasted with an image of unmoving tethered rats, through which the rats are displayed as suffering within the traditional experimental system. These images are allowed to speak for themselves, as Elspeth narrates to scientific audiences that telemetry reduces confounding variables, which is essential for conducting good scientific research. In the process, care shifts from a hidden to an explicit force that shapes the bodies of laboratory animals, whose biological development is ongoing and transmutable. By acting upon animals in a caring manner, Elspeth also argues that her research findings are more translatable in that the confounding variables that stress and suffering induce are reduced.

Expanding the Notion of Suffering

The tethered rats were not suffering in any way that is comparable to the dogs in Robert Hooke's experiment. Early humanitarians were responding to suffering in the context of what can only be described as physical torture. The 1876 Cruelty to Animals Act addressed this by requiring anesthesia to be used *if possible.* Emphasizing this clause is important because it highlights how experiments that were physically torturous for the animal could still be conducted, but under the condition that this be first approved bureaucratically by the Home Office (Holmes and Friese 2020; Shmuely 2023). The goal was to ensure that such research was only rarely conducted and only if the medical benefits were clear (Myelnikov 2024). And this continues to be the case today. The Home Office must approve all research involving animals in Britain, and all facilities conducting animal research must be approved and regularly inspected by the Home Office. The severity of the experiment and the amelioration of extreme suffering are important components of the application process. Nevertheless, the question of how to recognize and manage animal pain has been a key point of controversy in legislating (Myelnikov 2024) and practicing (Carbone 2024) animal research, with the veterinarian being the key professional who is responsible and accountable for this (Anderson and Hobson-West 2024).

In this context, I locate the moments of suffering I witnessed within a historical trajectory where suffering includes extreme physical pain but also exceeds it. The historian of science Robert G. W. Kirk (2014) has shown how the idea of "stress" transformed the ethics and eventual regulation of laboratory animal use in British science during the mid-20th century, expanding the scope of how an animal may suffer beyond physical pain to include social and mental distress. From the 1960s to the 1980s, the focus shifted from not only mitigating animal pain to also promoting animal well-being (Kirk 2014, 251). Using ethological knowledge, stress was defined according to how the animals related to one another in the cage and how they related to the scientists and animal technicians working with them. Kirk focuses on how the human is imbricated in the production of animal well-being in this context: "Stress made the physical and social environment determining factors of the physiological state of the laboratory animal under study. Furthermore, stress relocated the human subject within that environment, making the researcher integral to, controller of, and obligated to, the laboratory animals' well-being" (Kirk 2010, 258; see also Nelson 2018, 115–16 especially). Through this idea of stress, we can see the tethered rats as suffering, and this is also the logic through which Elspeth came to understand herself as being obligated to ameliorate that stress.

Elspeth addresses a suffering other within her conditions of constraint. She cannot change the lives of the rats that are before us, which she understands as a life of suffering. These rats are part of another, more senior scientist's experiment, and this experimental system has been approved by the Home Office. But she can change her own experimental system so that future rats will not suffer in this way. And she can try to convince other scientists to do the same.

Vignette

Several years later, in 2016, I conducted participant observation in the animal facility and some of the laboratories of a large research institute in the UK that I refer to simply as the Institute. The Institute has a large animal facility with two dedicated veterinarians. As a point of reference, Elspeth's laboratory and corresponding animal facility shared a veterinarian with other institutions. The Institute allowed me to witness the work of veterinarians whose job is to address and alleviate animal suffering. In contrast to the more haptic forms of care, which Elspeth introduced me to in her response to suffering, by watching veterinary work at the Institute I came to see how suffering is also viewed and ameliorated from a distance through surveillance technologies. As in the case of Elspeth, this care work is not done for the mice in front of the veterinarian but rather for future mice; the aim is similarly to avoid suffering.

I am spending the day with Vincent, one of the two veterinarians who work full-time overseeing the animal facility at the Institute. We have met and spoken many times before over coffee or as part of meetings, but Vincent notes that it will be rather unusual to have me shadow him for an entire day. He comments that the work of technicians is observable, but that he actually spends a good portion of his day in his office, at his computer. Vincent isn't sure what I will do or what I will see. I comment that I understand; I would struggle to show someone my work as well, given how much of it is spent looking at a computer screen. I didn't want him to feel that he had to perform for me; if he was at his computer, that was fine. As we neared the animal facility, and Vincent used his identity card to get us both into the building, he said to me, "You know Carrie, at any given time, we have about 30,000 mice in this facility. That is like a small village—a highly inbred, small village. I cannot be the heroic veterinarian, running from sick animal to sick animal here. I need to take a more public health approach and use data about the population to detect problems before we have a whole lot of sick animals."

The animal facility uses a computer system called the mouse information system, where technicians report the health of mice to the scientists and provide data for the veterinarians. Vincent gives me the example of teeth to demonstrate how he monitors the health of the mice from a population perspective using the mouse information system. Mouse incisors grow throughout their lifetime and are kept from growing back into the gums, in a circular fashion, by gnawing. Despite giving mice wood to chew on, dental malocclusion is a common health problem with laboratory mice. Here, a misalignment of the jaw means that the mouse's incisors do not occlude properly, and gnawing materials alone will not stop the incisors from growing back painfully into the gums. Traditionally, dental malocclusion is treated by trimming the mouse's incisors so the mouse can eat.

However, rather than focusing on individual treatment alone, Vincent also monitors the dental health of the population. Dental malocclusion is often the result of either trauma (e.g., food that is too hard, improper handling, fighting within the cage) or genetics (e.g., inherited through a mutation). Vincent will review both animal welfare protocols and breeding practices to see if any modifications can be made to avoid dental malocclusions. To be sure, Vincent is doing surveillance work, which includes surveying the work of animal technicians. Sociologists tend to view surveillance in a negative register as a site of power and control distinct from care. I want to emphasize that power and control are part of avoiding suffering and can be seen as caring. This finding detracts from the normative valences normally ascribed to surveillance technologies and care (see also Martin, Myers, and Viseu 2015). Caring across distance is, after all, a central feature of humanitarianism as well.

Preventing Suffering

The focus on promoting the health of laboratory animals was implemented over the second half of the 20th century through the professionalization of laboratory animal science as well as legislation. When the Cruelty to Animals Act of 1876 was updated with the Animals (Scientific Procedures) Act (ASPA) in 1986, the requirement of a Home Office license was maintained but with the added obligation that scientists adhere to the 3Rs—replacement, reduction, and refinement of animals in scientific research. The 3Rs is a concept developed by Russell and Burch ([1959]1967) in their *Principles of Humane Experimentation*, which aimed to make animal welfare concerns central to the conduct of science (Hobson-West 2009; Kirk 2018). The 3Rs require that science and scientists: (1) avoid or *replace* using animals in research by developing alternative models and tools; (2) *reduce* the number of animals in research by using the minimum number of animals statistically required through a focus on research design, and only using animals to truly add to existing knowledge; and (3) *refine* experiments to minimize the pain, suffering, distress, and lasting harm caused to animals as part of research (see www.nc3Rs.org). The goal of refinement was informed by the Five Freedoms of animals in the UK. Instituted in 1965 with a focus on agricultural animals, the Five Freedoms states that animals living under human control must be free to behave normally while also being free from: thirst and hunger, discomfort, pain, injury, disease, and fear or distress. The Five Freedoms provides a framework for understanding the tethered rats as suffering based on their inability to behave as rats because of the discomfort caused by the tether.

Multispecies Suffering

There is a genre of literature that fictionalizes the work of women in science, often authored by women who are scientists (e.g., Barbara Kingslover and Delia Owens). Gender and class are prominent themes, as are the relations between professional and amateur scientists, where natural history and biology are key points of reference. Yaa Gyasi's *Transcendent Kingdom* (2020) builds upon and diverges from this genre by making race and migration key themes, both of which she explores through the joint processes of producing knowledge in science and producing a career as a scientist. Her novel raises the vexed question of why scientists decide to pursue the research topics they do. In light of this, what kind of person is one expected to become performatively while producing knowledge with others, including laboratory animals? In the process, Gyasi's novel reorients the axes upon which science, religion, racism, and objectivity are discursively produced, and she creates a new narrative about science and humanism. Gyasi's novel reworks a largely white genre rooted in nature and natural history by placing the animal and the black female scientist in the laboratory. Gyasi thus critiques the operations of racism in science and does so in part by bringing the laboratory mouse to the fore as an actor.

Drawing on Bennett's (2020) literary method, I ask where and how laboratory mice act in Gyasi's novel as sites for troubling stereotypes regarding the roles of scientists and the norms of science. As a literary scholar, Bennett asks how black authors cultivate interspecies empathy to challenge how racialization and animalization are entwined. Drawing on Bennett, I have asked how Gyasi's novel critiques Western philosophical thought that has entwined animalization and racialization while also articulating interspecies relations in new ways. Her novel reflects ways out of two polarization cycles, firstly of animal use and abuse but also between religion and science.

Transcendent Kingdom begins with the protagonist Gifty recounting her image of her mother's depression. She describes it as a queen-sized bed that her mother colonizes "like a virus" for months on end. Her mother's first episode of depression occurred when Gifty was a child, and so she was sent from the United States to Ghana to wait out her mother's depression with family. Her mother's second episode of depression occurs while Gifty is completing her PhD in behavioral genetics at Stanford. It forms the present of the novel—a present that connects Gifty's childhood to her future. Gifty receives the call from her mother's pastor; the depression is happening again, and this time Gifty's mother is being sent from Alabama to Stanford to wait the depression out. The phone call arrives while Gifty works with her mice in her lab. She is not conducting what is often considered "science" at this moment, but rather the care work that is frequently central to a PhD student's experience of doing science. Gifty is caring for her mice, trying to separate them after a bad fight resulting in one of them being injured. After she picks up her mother from the airport, brings her mother to her apartment, and hears that her mother is asleep in her bed, Gifty returns to the lab to check on her mice.

> As soon as I heard the sound of soft snoring, I sneaked out of the apartment and went to check on my mice. Though I had separated them, the one with the largest wounds was hunched over from pain in the corner of the box. Watching him, I wasn't sure he would live much longer. It filled me with an inexplicable sorrow, and when my lab mate, Han, found me twenty minutes later, crying in the corner of the room, I knew I would be too mortified to admit that the thought of a mouse's death was the cause of my tears. "Bad date," I told Han. A look of horror passed over his face as he mustered up a few pitiful words of comfort, and I could imagine what he was thinking: *I went into the hard sciences so that I wouldn't have to be around emotional women.* (Gyasi 2020, 9)

Gifty's relationship with the mice is "demanding" (Munro 2004), and she cares about and for the mice and learns from them. But this is certainly not a relationship where mice are humanized literally or symbolically.

> I watched my mice groggily spring back to life, recovering from the anesthesia and woozy from the painkillers I'd given them. I'd injected a virus into the nucleus accumbens and implanted a lens in their brains so that I could see their neurons firing as I ran my experiments. I sometimes wondered if they noticed the added weight they carried on their heads, but I tried not to think thoughts like that, tried not to humanize them, because I worried it would make it harder for me to do my work. (Gyasi 2020, 20–21)

Mice and humans are and remain different here. Gifty and her mice become together, "alongside" (Latimer 2013a) one another, without collapsing into one another. They are in relation such that interspecies empathy occurs, but without animalization. In defining "being alongside" conceptually, Latimer emphasizes that this is a difference-preserving process. She states: "Preserving division—a conjoining of contingent and different 'parts,' none of which is simply subsumed into a whole. For example, being alongside can involve cooperating with one another, even working together, but not with the same materials, and not necessarily to the same ends" (Latimer 2013a, 79–80). Gifty recognizes that she is inflicting suffering, despite anesthesia and painkillers, because of the weight that these mice are carrying in order to have their brains visualized.

In this passage, the readers see Gifty try to deny the suffering of the mouse by not thinking about it. This denial is part of what extends Gifty throughout the novel and multiplies her. The narrative structure of *Transcendent Kingdom* enacts the psychological model of trauma, wherein an exiled childhood self is forcing her way into Gifty's self-system, across the novel—as it moves between present-past and present-future.

> I started to feel like I didn't have a self to get a hold of, or rather that I had a million selves, too many to gather. One was in the bathroom, playing a role; another, in the lab staring at my wounded mouse, an animal about whom I felt nothing at all, yet whose pain had reduced me somehow. Or multiplied me. Another self was still thinking about my mother. (Gyasi 2020, 10)

Gifty's mother's second episode with depression reopens the past as Gifty finds herself caring for her mother while completing her PhD. Alongside the mice, Gifty's mother plays a key role in forcing Gifty to work through her past-present, in part highlighting what formed her interest in behavioral genetics and her personhood as an unemotional, scientific woman. Through these reflections, the specters of two other, now absent, family members surface: Gifty's father and brother.

It is worth considering that injured mouse that multiplies Gifty as she is forced to reconcile her childhood self—which includes being devoutly religious—with her adulthood and very much scientific self. The recovery model of trauma, which Gifty's story enacts, is rooted in addressing suffering. The mouse in pain makes Gifty grieve for her own suffering, and her mother's suffering, which she previously denied. This move in the novel aligns with how "being alongside," as a concept, troubles the notion of the individual by making not only visible but also integral the relations that make up any self (Latimer 2008). Gifty becomes by making those relations visible to herself, in a process that Latimer, following Strathern, refers to as a relational "dividual" as opposed to a bounded, independent "individual."

As the novel develops, we, the reader, slowly learn of the two traumas in Gifty's childhood. The racism in the United States is simply too excruciating for Gifty's father, whom she refers to as the Chin Chin man; he cannot be a man in the United States and returns to Ghana, leaving his family in Alabama. This trauma shapes the people around Gifty more than Gifty herself, as she is too young to remember his presence. Second, Gifty's brother had been a talented basketball player as a teenager, so much so that he seemed to alleviate the racism that pervaded their lives by being a star athlete. But after being prescribed oxycotin following an injury, her brother becomes addicted to heroin and, following several relapses and much family turmoil, dies of an overdose.

While the novel follows a psychological model of trauma rooted in the internal family system approach, in sociological terms Gifty must reconcile the part of her who is a scientist today with the part of her who was devoutly religious as a girl until her brother died. Following her brother's death, Gifty's suffering and her mother's suffering were compounded by their ostracization. Gifty's brother's addiction was seen as a racialized moral failure rather than a deeply unjust form of systematic political and economic exploitation and racism. A postdoc in Gifty's lab named Katherine, who had practiced psychiatry before doing a PhD in neuroscience, extends a hand of friendship to Gifty throughout the novel—persisting despite Gifty's resistances—in telling this psychologically informed story of trauma. Gifty, meanwhile, gives voice to the sociological dimensions of her suffering, as marked by migration and racism. Gyasi makes palpable for the reader what it feels like to be a black woman in a field of science dominated by white men, emphasizing how this inequity overdetermines without determining Gifty's subjectivity.

Gifty decides to study neuroscience in part to address her own suffering. She wants to understand addiction and depression to understand her brother and mother. She makes mice suffer as part of this project, which will not take away her suffering or her mother's suffering. Her research will not bring her much-loved brother back. But Gifty hopes that her research, and all the human and other-than-human suffering that goes into it, might, someday, help someone else.

More-than-Human Humanitarianism and Suffering

Today, psychotherapy is drawing extensively upon pan-spiritual teachings that argue suffering is something that needs to be accepted and gone through. We see this combination of psychotherapeutic and spiritual thought in Gyasi's novel. Gifty cannot deny that she was making the mice suffer as she begins to accept her own suffering, as well as her brother's and mother's suffering. But in contrast with Victorian-era Christians, this does not mean that Gifty rejects the use of animals in her research. She instead accepts that she cares about the mice, not just because she needs them for her research as utilitarianism would emphasize but also because she wants to reduce their suffering in her encounters with them. She quite simply cares.

What humanitarianism and more-than-humanitarianism share, from the vantage point of Britain, is a desire to end suffering. It is difficult for me to read about Hooke's experiment without feeling sickened. And yet, I have ingested the lives and deaths of many laboratory animals through my consumption of biomedicine. The logic of the animal model paradigm is that (some) human health is improved through the lives and deaths of animals who likely suffered. That said, I think that the suffering of the mice and rats, presented in both my fieldnotes and in Gyasi's fictive account, is qualitatively different from the torturous experiments conducted by Hooke. Animals suffer in science, but they do suffer less because people have not only sought to abolish their use but have also ameliorated their suffering in the doing of science.

Accepting suffering was itself central to antivivisectionism in 19th-century Britain. Antivivisectionists believed that suffering was linked with redemption rooted in Christianity. In this context, suffering cannot be eliminated, avoided, or denied; it must be accepted. In this sense, it is worth remembering that animals do not need to suffer for human health if "we" decide to accept our lack of physical and/or psychic ease, or disease. From my vantage point, we seem very far from this type of Victorian ideal. Moreover, the degree of moralism and attendant victim blaming within that Victorian ideal should give pause. Nevertheless, the vast use of animals in creating and alleviating human illness could be intervened in by focusing more on the social production of health rather than the biomedical treatment of disease.

Albeit with a different grounding, feminist science studies rooted in Donna Haraway's (2008) "shared suffering" also explores this idea that going through suffering, relationally, can further delimit it. In the context of laboratory animals specifically, Gail Davies (2012a) has argued that an acceptance of suffering among laboratory animals, rather than its denial through universal denunciation, could work to make the particularities of suffering—a dog splayed open, rats tethered, mice surgically instrumented—unforgettable, and thus undeniable. By seeing the tethered rats as suffering, rather than denying this through recourse to bureaucratic justifications or "trying not to think about it," Elspeth experiences suffering as well. She is motivated to alleviate, if not these rats' suffering, future rats' suffering, in light of her experience of shared suffering. She develops an alternative experimental system as a result.

In contrast to a utilitarian point of view that accepts suffering if the benefit outweighs its cost, seeing suffering and acknowledging its existence rather than denying it can compel new ways of relating to suffering and its alleviation. These are not idealistic solutions that will abolish suffering. The mice are, after all, still instrumented in Elspeth's experiments. As Gifty notes, being instrumented with telemetry is probably, at the very least, uncomfortable. Further, mice will die to produce knowledge that might help some human health but will certainly not help the mice. The imperfect work of more-than-human humanitarianism is a mutation from the Victorian-era, antivivisectionist ideal.

Humanitarianism through the Prism

What we see of humanitarianism through this lens of the laboratory animal is that the collapse of humanitarianism into human rights is not self-evident. If humanitarianism is about saving lives in a crisis situation as an outgrowth of charity, human rights have a longer time horizon and a more political orientation rooted in addressing the root causes of suffering (Krause 2014, 149). This distinction has been important for humanitarianism, as neutrality has historically been a precondition for humanitarian aid workers being able to gain entrée into conflict regions to provide assistance. The politicization of humanitarianism, both within relief organizations and by nation-states and other political actors, has challenged the idea of neutrality and the ability to provide aid. Indeed, the idea of neutral and disinterested knowledge is not tenable in a purist sense. But what I think more-than-human humanitarianism shows is that these tensions and limitations do not mean that humanitarianism and human rights must hybridize.

More-than-human humanitarianism is, in contrast, distinct from animal rights in an, at times, highly conflictual but often productive relation. The two have not hybridized but remain only ever in partial connection. People involved in animal welfare are well aware of how their work can legitimize animal use, and scientists may welcome the animal rights activists' interventions for how they have challenged scientists to reduce animal suffering.

I will give an example of how difference works in important ways within more-than-human humanitarianism. Larry Carbone, a veterinarian, asserts in the *Researching Animal Research* (2024) edited collection that having outsiders witness work in animal facilities is important. Carbone states:

> Vets, like patients, are at risk of being propaganda tools for animal research, at risk of performing the role of the human carer and healer. But this carries risks, I believe, for the actual care the animal receives. . . . I see facial/grimace scoring as the NVS's valiant effort to make up for the fact that vets in labs do not (and in the current scales of efficiency in laboratory work, cannot) devote the same time and effort into animal patients' pain management as companion animals. . . . Outsiders may make for better quality science and better treatment of animals. (Carbone 2024, 317–18)

Carbone is worried that standardized pain management tools—like facial grimace scores—improve care only within the context of caring for a large population of animals. He worries about the reliance of laboratory animal vets on surveillance rather than haptic forms of care, as seen in Vincent's work. Through the work of outsiders, in this case social scientists, Carbone sees the political economy of laboratory science and that this political economy does not allow for more time to be spent with animals, which would allow for better pain assessments. More time with animals could improve care, and inviting outsiders into the laboratory helped Carbone as a veterinarian to see this as a structural condition. The Animal Research Nexus enacts the approach of being alongside, where being in relation is an extension rather than a hybridization that aims at doing science with animals better (G. Davies et al. 2024).

This raises for human humanitarianism and human rights the possibility of disentangling humanitarianism and human rights and questioning how the two can be put productively alongside one another in order to address suffering at a range of scales and temporalities. What would humanitarianism and human rights look like if they were not combined, and it were easier to see where one ends and the other begins? What if humanitarianism and human rights were in a more antagonistic relationship with one another? Both humanitarianism and more-than-human humanitarianism are uniquely in danger of being coopted by powerful institutions, including nation-states and biomedicine. This danger needs to be critiqued by outsiders. Distance and difference can be an aid in such critical practices. I do not want to romanticize this, however. Distance can descend into violence, as anyone who works in the area of laboratory animals in Britain knows. But critical distance can also allow for different forms of "incommensurable care" (Giraud 2024) to be helpfully "choreographed" (Thompson 2013, 2005). I would like to suggest that this is an important insight for any area of social life that is stuck in the polarization cycle.

2

Care

Over time, Elspeth, the scientist who originally introduced me to laboratory animals and allowed me to watch her laboratory's work, shifted her research to data science and stopped conducting animal experiments herself. I could not continue to observe how animal care was part of her science anymore because she was caring for animals by not using them in her research. In this context, another scientist and interlocutor introduced me to what I will call the Institute. He thought that the Institute would be an interesting place for my research on care because of the longevity of the mice in their animal facility: the mice were living far longer here than at any other facility that he was aware of. Longevity was seen to be proof of high-quality care. Within the Institute, I met Vincent, one of the two veterinarians based there.

How Elspeth had seen suffering and responded to that suffering with care was different from how the Institute saw suffering and responded. Elspeth used her body and our physical proximity to rats and mice in order to show me suffering and her response to that suffering. Her work was rooted in haptic forms of care. In contrast, Vincent's approach to seeing suffering from a distance was interlinked with his approach to doing care work from a distance. On my first visit to the Institute, I was given a tour of the animal facility using CCTV. I was told that we would need to schedule a specified time in advance and full days for me to visit the inside of the animal facility. Because it was a biosecure facility, everyone had to take either a wet or dry shower before entering in order to avoid being a vector for contamination. It was different from my experience in Elspeth's laboratory, where no showers were required and scientists worked side by side with technicians. The Institute allowed me to see how care from a distance operates in relationship to haptic care, which is more often the focus of empirical studies (see Puig de la Bellacasa 2017).

This chapter explores care practices that are often rendered invisible in science, wherein haptic care and care from a distance occur in tandem in relationally produced knowledge that goes beyond the human.[1] In doing so, I draw on the long-standing argument that care is not only a practice but also a form of knowledge, albeit one that is marginalized. Much of the research on the crisis of care in the current moment is rooted in political economy (e.g., Collective et al. 2020) and traces back to the work of Joan Tronto (1993). The focus is on critiquing the devaluation of care work. I find this work particularly conducive to thought. Still, I argue that to understand care work in the context of laboratory science, I have found it more useful to root care in knowledge and to trace this back to the alternative lineage of Carol Gilligan (1982). I present here care as a form of embodied knowledge that requires regular practice to sustain it.

Care as Work and Knowledge

To introduce care as both work and knowledge, I refer to my observations with Janet, who was caring for very old female BALB/c mice at the Institute. One of the things I learned during my time with Janet in 2016 was that she was trying to improve the lives of these elderly mice above and beyond the legally required care. She did care work supplementary to what the veterinarians had decided represents best practice. Janet explained to me her concerns over the well-being of these mice, who live in small cages for such a long time, which she believed must be rather boring. The Institute used sunflower seeds as a form of enrichment for its mice, and her concern was that sunflower seeds were too easy to find and open. Janet worried that the sunflower seeds would not keep the mice properly interested, especially given that these mice were kept alive for so long.[2] She told me that she thought mustard seeds, which are smaller and more difficult to find and open, might be more enriching for these very elderly mice. Given the biosecurity requirements of the facility, Janet needed to find a supplier who could package mustard seeds so that they could be irradiated before entry to the building. Janet had been spending her spare time trying to find such a supplier.

Astrid Schrader (2015) has distinguished between "caring for" and "caring about" (see also Tronto 1993). Schrader notes that caring for is goal-oriented, and the receiver of care is defined by a lack of ability or autonomy. Caring about, conversely, does not have a predefined objective for care but is an affective relation open to becoming in a more open-ended manner with another being, human or other-than-human. Caring about can include those whose existence one may have not been previously aware of (see also Puig de la Bellacasa 2015, 2011; Latimer and Puig de la Bellacasa 2013; Haraway 2008; Despret 2008, 2004, 2005; Latimer 2013a, 2011). Janet showed me the different kinds of care she practiced as she moved between caring for and caring about laboratory mice. She showed me that she *cared for* laboratory mice very well and according to veterinary knowledge and the science of animal welfare. Being an animal technician is, without a doubt, a caring profession. But Janet also wanted me to understand that she *cared about* her mice, so much so that she had come to care about mustard seeds as well. This practice requires caring about things that no one else in the workplace may even think of, like sunflower seeds, mustard seeds, and product packaging.

Care work has historically been unpaid work. It has been women's work to biologically and socially reproduce the species and thus create a human workforce and a consumer force. Care work has been naturalized and devalued when framed as inherent to biology and instinct; it is described through the language of hormones and desires that connect individuals with populations through evolutionary need. Where a wage is introduced, it may establish a (usually low) value for that work, but it also seeks to disconnect or sever the caring work we do from our person. The care no longer belongs to us but rather to our employer, and the fact that we care becomes a way to extract more labor from us at a lower wage. It is alienation, and wherever there is alienation, there is replaceability. The long-standing, political economic critique of care and its systematic devaluation through capitalism, patriarchy, and racism highlights the interchangeability of self-contained units as the basis of labor under capitalism.[3]

These structural critiques are important, but it seems to me that they also risk neglecting many elements of care. In response, I ask what gets lost by focusing solely on critiquing Janet's labor as unpaid when she learns about mustard seed distribution. In contrast, I have learned a lot about knowledge and the importance of valuing different knowledge from Janet and her unpaid work. Indeed, Maria Puig de la Bellacasa's (2011) argument that care is a neglected but world-making practice, an affective state and a material way of doing, has provided an important lens for science studies to explore care in science as practice. Building on this tradition of care studies from within STS, I want to emphasize the knowing elements of care. What happens to care when we see it not only as work, not only as obligation, but also as a site of knowing—one that needs to be described and valued?

The ethic of care, as developed by American psychologist and ethicist Carol Gilligan in her 1982 study *In a Different Voice*, emphasizes that care is a way of knowing. Gilligan argued that men were more likely to reason according to universal ideals rooted in justice, while women were more likely to reason according to relational ideals rooted in care. Gilligan located these types of moral reasoning in gendered and gendering socialization processes. The phenomenon could be seen in both children's socialization processes and in psychologists' evaluations. In other words, Gilligan emphasized both the interactional and the structural in showing how knowledge practices were gendered in ways that reproduced patriarchy. Girls were taught to think relationally at the interactional level, and this form of knowledge was systematically and structurally deemed inferior within the knowledge practices of psychology.

Within the knowledge practices of psychology, Gilligan identified the marginalization of relational ways of knowing, a marginalization that was embroiled in the reproduction of sexism in the 1980s. What can we learn about social life today through the marginalization of relational ways of knowing in the knowledge practices of the life sciences? What we see is that while care may increasingly be understood as a key part of science, its micropractices are still nonetheless at continual risk of being rendered invisible.[4] Rendering the invisible visible is a key goal of this book, because to truly value care its operations need to be described and represented.

Vignette

Today, I am shadowing Martine—a young animal technician who has recently switched her focus to laboratory animals. Martine says that she enjoys this work and enjoys science. She immediately tells me that she finds my work as a sociologist—to study the social life of people as a scientist—confusing. Martine proudly presents herself as an animal person, and as someone who would much rather be in the presence of animals. She longs for a cat, not a baby—much to her mother's concern. Martine finds me, or rather a sociologist, a curiosity that she is nonetheless intrigued to learn a little more about during our day together. The fact that I am also the mother of a human child is of little interest to Martine; we instead talk a lot about my dog. In various ways, animals are the point at which Martine and I meet one another, making it possible for us to "be alongside" (Latimer 2013a) one another and to find moments of partial understanding (see also Latimer and Lopez Gomez 2019b). The intimacy of the home—pets, mothers, and children—grounds our conversations about what it is like to care for laboratory animals.

Martine explains that today, Friday, is a slow day. All she will do is check the cages that are the homes of the mice she cares for. The cages are checked twice daily, per Home Office regulations of laboratory animal welfare practices in the UK. But Martine explains that the Friday-afternoon check is more rigorous. If she thinks any cage may be at all low on food or needs a change of bedding, Martine wants to address this now. Doing this work now is as much for the mice as it is for her colleagues. The animal technician who will be coming in over the weekend to check on the mice twice per day can do so rather quickly if the Friday-afternoon check is thorough. Martine also doesn't want any risk of the mice not having their needs met over the weekend.

Martine begins by ensuring the water filtration system works for the rack, which holds approximately 100 cages containing anywhere from one to six mice. She then pulls out one cage at a time to visually check the mice, ideally without opening the cage and disturbing it. If Martine needs to open the cage to give more food or to check on the mice more closely, she will flip the tag on the cage upward so that she can do this additional work after checking the entire rack.

I comment that Martine clearly sees things I cannot see as she checks the mice without opening their cages.[5] Martine agrees with my assessment but struggles to put into words exactly how she goes about *knowing* that the mice are okay by looking into, but not opening the cage. A few minutes later, she says, "Do you want me to show you a trick?" I eagerly say yes. Martine is holding a cage in which all the mice are huddled under a red tunnel, which I have been told allows them to feel like they are in the dark while the animal technician can still see them inside the cage. Martine can then count the mice and visually ensure the correct number of mice are present in the cage. Martine has the cage rested in one hand, with the other hand on the side of the cage, and is holding it at eye level so that she can look in. She tells me that she always puts her hand underneath the mice on the bottom of the plastic cage to ensure that she *feels* their body heat. She knows the mice are alive and well by feeling their warmth through the plastic cage. That way, Martine checks that they are physically okay by feeling the traces of their presence, but she does not have to disrupt the mice and wake them up by opening the cage. She reminds me that mice are nocturnal and like to sleep when we want to work.

I ask Martine if someone taught her this trick, as she has just taught me. She says no, it is just something she picked up by herself over time. Martine recalls that she hadn't realized she was even doing this per se; she just always put her hand on the part of the cage underneath the space where the mice had nested themselves and felt their body heat through the plastic. I interpret Martine as saying that it is instinctive to meet the animal through the heat that their bodies produce. Indeed, it is comforting for humans to feel the warmth of animals; I seek out the warmth of my dog. The anthropologist Hanna Kienzler has told me people will often build sleeping quarters above animal barns to feel the warmth of the animals below. It is always both utilitarian (to stay warm) and pleasurable (it is comforting and nice to feel the warmth of an animal).[6] Heat is an "image" of well-being, a suggestive truth that can "match the density of our feelings" (Stevenson 2014, 13). Martine continues to tell me that one day she didn't feel warmth when she put her hand under the nested mice and immediately knew something was wrong. She comments that it was probably at this point that she became aware of what she had been doing all along when she checked the mice in the cages, holding the cage in her hand under the place where the mice were huddled within the red dome.

Martine pauses and reflects for a minute. She then sums up her work: the opposite of a physician's work. She goes on to explain that she has a sense of what is normal by working with hundreds of cages of mice, day in and day out, week after week and month after month. When things aren't normal, she continues, she seeks to figure out what is wrong. Martine then tells me this is the opposite of the doctor; the doctor works with hundreds of sick human patients to learn about abnormalities. "Physicians start with abnormality; I start with normality," she says.

Care and More-than-Human Knowing

I want to explore Martine's sense of heat in mouse cages through Eduardo Kohn's (2013) theorization of knowing beyond the human. Kohn develops this theory vis-à-vis the work of Charles Sanders Peirce, and counters the idea of the mind as something individualized and limited to a brain. Rather, mind is something that is relationally produced. Peirce, and much of symbolic-interactionist and pragmatist thought, locate this relational mind in interactions between humans; Kohn, however, extends this model to include other-than-human interactions. Relational models of becoming that predominate in STS are brought together with the animist thought of the people of Ávila, Ecuador, with whom Kohn is conducting ethnographic research. Kohn's work makes explicit the conceptual links between new materialist, feminist thought in STS and the animist knowledge practices of indigenous thinkers and people from around the globe (see also TallBear 2017 for this critique).

An example is helpful to explain Kohn's theory of more-than-human thought. He demonstrates his argument by describing an encounter between humans and monkeys. Kohn (2013, 30) recounts being in the forest with a man named Hilario and his son Lucio, where they came upon a troop of woolly monkeys. Lucio shot and killed one monkey, resulting in all but one young monkey dispersing. The young monkey was hiding in the canopy, and Hilario and his son tried to make it move in order to shoot it. Hilario started cutting down a nearby tree, saying things like "look out" and making word-like sounds that mirrored cutting the tree, such as "ta." Kohn argues that the crashing tree stood for something to the young monkey, a sign of something "dangerously different" (2013, 35). So the young monkey jumped—as Hilario was hoping. Kohn argues that humans and monkeys interpret "indications" or "indices" that connect two events—a crashing tree and danger—and thus communicate without shared objectives or language. Signs (using Peirce's definition) are made and interpreted across species and within this specific encounter. Previous encounters provide a means to do that interpretive work.

Through Kohn's work, I came to understand the heat that mice produce, through their bodies by being close together under a red dome, as an "indication" or "sign" (Peirce [1894]1998) of wellness. The warmth produced by bodies is an "index" in Peirce's sense: "Anything which focuses the attention is an indication. Anything which startles us is an indication, in so far as it marks the junction between two portions of experience" (Peirce [1894]1998, 8). The "habit of life" (Kohn 2013) to produce heat, store heat, and seek out heat is part of a generality of thought not contained in a singular mind–body in the Cartesian sense. Rather, it is distributed across bodies that think. Martine could only become aware of this habit when it was disrupted, when she did not feel the heat at the bottom of the cage and was thus startled into knowing something was wrong. These interactions produce knowledge in the form of a more-than-human and relational mind as well as produce the materiality of the world.

Kohn, therefore, makes a crucial point that we should not conflate representation with language, nor should we assume that representation is a distinctly human activity rooted in Saussure's model of the arbitrary sign and signified. Language is one mode of representation, among others. A far more common mode of sign processes is to take something as significant and to act in response to what that sign stands for in a manner that will, in turn, be a sign to all the other life forms one is entangled with. It is a classically pragmatist formulation that links the interactional with the structural by foregrounding the role of historical experience in interpreting and acting in the here and now. Kohn extends this model of social life to be a model of knowledge.

Disruption and shock are crucial processes in this way of knowing. For Martine to make explicit her knowledge that she was seeking out heat as a sign of wellness, she had to experience the shock of not feeling the heat. Disruption and shock rely upon a prior indistinction and the accumulation of knowledge that is implicit. Kohn highlights the idea of indistinction by describing the insect known as a walking stick. He starts with the usual understanding of icons, where two things known to be different have similarities: a stick and an insect that looks like a stick. However, iconicity requires a prior step of *not* noticing a difference. For the insect known as walking sticks to evolve, its ancestors had to *not* be noticed by predators as prey. The relationship between heat and wellness became an indexical reference for Martine when it was missing, making it possible for her to know that something must be wrong when noticing a change. The knowledge that this shock produced relied upon a long period of habituation to indistinction.

The monkeys in Kohn's story were habituated to the normal and therefore knew that their current situation was not normal and, therefore, dangerous. And they knew to flee. The young monkey was not yet habituated in this way, however, and did not flee as the rest of the troop had; instead it froze. Similarly, there was a moment for Martine when she realized her mice were not in a normal situation, and quite possibly she froze as she worked out what to do. Martine put her indexical knowledge, produced with the mice and cage, into language or her second language, English, in order to convey this knowledge to me. Heat and wellness became a symbol at this moment, or a fact rooted in an essential and ontological relation (see C. L. Briggs 2007, 324). Putting this knowledge into language is simply one moment, however, in producing this relational mind.

We can see mice as living beings, or selves in Kohn's more-than-human theory of knowledge, who make nonsymbolic signs within the animal facility with one another, with cages and enrichment toys, and with humans.[7] The animal facility is an emplaced semiotic field where mind is relationally produced across species and things. Unlike the Amazon in Ecuador where Kohn did his fieldwork, the animal facility is premised upon being a highly restricted ecology of selves. Nonetheless, care and knowing become almost indecipherable here, as both depend upon the attachments between Martine, the mice, and the cage. Knowing in the animal facility will never be complete because it requires so many indexical relations to go unnoticed. Not noticing is not a problem, but rather a condition of knowledge. As one of the Named Animal Care and Welfare Officers (NACWOs) at the facility commented, "The more we seem to know about mice through this facility, the more questions we have." This is because everyone is coming into being together and in the flow of life, which will also always generate surprises. Martine's care work is thus knowledge work that makes the materiality of the world.

Transcendent Science

I now amplify this vignette by exploring themes in Gyasi's novel *Transcendent Kingdom* that relate specifically to the topic of care. There are many reasons why Gifty becomes a scientist. She was a talented young person in school, and wanted to do "the hardest thing you could do" (Gyasi 2020, 33). She becomes a scientist because she wants to understand transcendence and, as she admits across the book, is responding with science to her brother's addiction and untimely death. Gifty must accept that she cares about addiction and its amelioration, and this care is something that she struggles to make space for in her scientific self.

Transcendent Kingdom opens with the grand narrative of science, wherein mice as models can stand for humans. The brain is presented as an organ, thought to be the locus of self and personhood. Gifty states:

> Though I had done this millions of times, it still awed me to see a brain. To know that if I could only understand this little organ inside this one tiny mouse, that understanding still wouldn't speak to the full intricacy of the comparable organ inside my own head. And yet I had to try to understand, to extrapolate from that limited understanding in order to apply it to those of us who made up the species Homo sapiens, the most complex animal, the only animal who believed he had transcended his Kingdom, as one of my high school biology teachers used to say. That belief, that transcendence, was held within this organ itself. Infinite, unknowable, soulful, perhaps even magical. I had traded the Pentecostalism of my childhood for this new religion, this new quest, knowing that I would never fully know. (Gyasi 2020, 18)

Gifty is thus inspired by the Cartesian mind, where the brain is the source of knowledge. STS has sought to trouble this idea of mind and show it as one very productive but incomplete way of theorizing life itself.

By the novel's end, however, this grand narrative is replaced, and we see Gifty articulate a different relationship with science, one that expresses her everyday practices of working with mice throughout the novel. "My work pursuits are much more modest: neurons and proteins and mammals. I'm no longer interested in other worlds or spiritual planes. I've seen enough in a mouse to understand transcendence, holiness, redemption. In people, I've seen even more" (Gyasi 2020, 246). It is not a story of heroic science but of a very uncertain science; a science that may or may not result in interventions that would improve the lives of humans, of loved ones.

"You're doing real good," I said to the mouse as I put him down. Though I'd repeated this process dozens of times without fail, I still always said a little prayer, a small plea that it would work. . . . Many, many years down the line, once we've figured out a way to identify and isolate the parts of the brain that are involved in illnesses, once we've jumped all the necessary hurdles to making this research useful to animals other than mice, could this science work on the people who need it the most? Could it get a brother to set down a needle? Could it get a mother out of bed? (Gyasi 2020, 40)

The humble aspirations of Gifty's science do not mean she stops doing the research. Gifty's work continues, and she continues to use mice to understand humans in ways that could benefit future people like her brother and mother. The grand narrative of science has been replaced with a more humble and tentative set of scientific aspirations that are *also* rooted in, I would say, care.[8]

Gifty and the mice become alongside one another, meeting "in connection but clearly doing different things" (Latimer 2013a, 95). Latimer is centrally concerned with how relationalities are productive when different beings are alongside one another, quite often with different goals. This differs from Haraway's merging of different beings in companion species. Haraway is a biologist turned social theorist who wants to understand how bodies of all kinds are relationed forth. In contrast, Latimer is a nurse turned social theorist who wants to understand how relations are bodied forth. I see Gifty and her mice as fitting clearly within Latimer's being alongside, articulating the differences that are bodied forth when a togetherness is not necessarily rooted in companionship.

Being alongside is compatible with Bennett's conceptualization of "becoming alongside" (Bennett 2020, 132). He articulates this interspecies ethic through a reading of Zora Neale Hurston's corpus generally and *Their Eyes Were Watching God* specifically. Bennett emphasizes relations that are fleeting and potentially fraught, which hinge "on one's willingness to pay attention to the flesh—to care for it, even and especially when that flesh is not held precious by the protocols and practices of Man and the human as it has been historically imagined within the Western philosophical tradition" (Bennett 2020, 135). Bennett turns to Édouard Glissant and the black feminist ethic of care in critical dialogue with Heidegger. Transcendence of self comes from an ability to be alongside another and not deny the suffering that those relations body forth. In becoming alongside, we experience transcendence or the becoming of more than ourselves as "dividuals."

Through Latimer and Bennett, I have come to understand Gifty as being and becoming alongside the mice, connected but different, in a manner that radically diverges from the ideas of similarity and difference articulated through the model-organism logic itself. According to that logic, mice stand in for humans as surrogate bodies that can be researched upon in ways that humans cannot be. The mice and the fact that Gifty cares about those mice is not a matter of substitution, however, but rather of relations. Gifty's relations bring forth the fact that she is a scientist in part because research provides a way to mourn her brother's death, confront the racism that contributed to his death, and address the fact that she lost both God and her mother through her brother's death. Gifty is a scientist because she cares, and that fact surfaces through her relations with both mice and people. Gyasi's book ultimately asks why a scientist must repress so much that she cares about for her knowledge to be deemed legitimate.

More-than-Human Humanitarianism and Care

Relational modes of knowing are infrastructural to a science that gains its epistemic authority by being universal. What happens to knowledge if we make those practices visible rather than deny and repress care? Gilligan explored how universal versus relational modes of knowing both instantiated and reproduced gendered hierarchies in the 1980s. She wanted to place value on relational knowledge practices, which women were more likely to engage in. Today, relational modes of knowing continue to be fought for. Feminist science studies scholars including Donna Haraway (2008, 1991), Karen Barad (2007), and Vinciane Despret (2004, 2008, 2013), in addition to anthropologist Marilyn Strathern and sociologist Joanna Latimer, have all been crucial in putting forward relational models as not only a different but also a better way of knowing the world. The goal here isn't to change the demographics of knowledge producers, as liberal feminism would do, but rather to change the epistemological and ontological base of knowledge itself. Kim TallBear (2017) has meanwhile shown that these relational and more-than-human theories align conceptually with much of indigenous thought from various parts of the world but that these indigenous thinkers are rarely referenced within these science studies texts. There are layers of neglect when care is understood as knowledge and as marginalized knowledge.

Tone Druglitrø (2018, 658) notes that, during the mid-20th century, the work of "animal caretakers" in science was renamed "animal technician"—the nomenclature that this book uses and thus reproduces. She notes the irony that as care work was being more clearly defined and respected as a crucial part of doing "good science," the word "care" disappeared. In delineating this shift, Druglitrø cites a Norwegian veterinarian who stated that "this is no longer a place for people who are unsuitable for everything else" (Druglitrø 2018, 658). Care was thus demeaned as something that only people who could do nothing else engaged in; technology was a means of professionalizing the workforce. But Druglitrø argues that this did not mean that care disappeared; an ability to engage with animals affectionately remained part of the job (see also Holmberg 2011). This remains true today. Greenhough and Roe (2019) note that animal technicians who do not themselves have pets are looked upon with some suspicion. Care is a crucial but generally invisible aspect of scientific work, as a form of tacit knowledge (Holmberg 2011) that is often "forgotten" or "strategically ignored" (G. Davies 2012a, 7).

But why should all the careful knowledge of animal technicians and scientists—the private cares about mustard seed distribution or how heat on the bottom of the cage can be an indicator of wellness to those who are caring for mice—be repressed?

Gilligan's work on the ethic of care is helpful to return to here. First, the ethic of care has been taken up to study the relationship between gender and knowledge regarding veterinary medicine. Veterinary medicine has been marked by a rapid shift in the number of female practitioners since the 1980s, as a male-dominated profession becomes a female-dominated one.[9] The increasing number of women in veterinary medicine has raised questions about the extent to which changing the gender profile of the profession will change its knowledge practices. Two perspectives have emerged in this context; one argues that women veterinarians are transforming the profession, and the other claims that the profession remains masculine even though women are becoming more prominent.[10] In other words, the feminization of veterinary medicine does not necessarily mean that more relational ways of knowing will come to the fore—and indeed, this has been a key argument of feminist science studies scholars like Judy Wajcman (2004, 1991), who has forcefully argued that bringing more women into science will not necessarily change its ethos in the ways liberal feminism suggests.

Drawing on this strand of ethics of care scholarship, I would argue that neither Janet's nor Martine's knowledge is repressed primarily because they are women. Rather, their knowledge is repressed because it does not fit a science that has a masculine ethos, one that has developed over the *longue durée*. This scientific ethos still sees the material world as an objective and distinct reality and thus seeks to cancel out relations through the promise of standardization. Science has a goal: to standardize care in the form of a pharmaceutical rather than to body forth the world through caring relations. This version of science can understand the laboratory animal as an "object of care" (Druglitrø 2018), but struggles to see the relationally produced mind of the animal facility that is produced across bodies and species.

Oppression in the animal facility, I would suggest, therefore, involves repressing the relational mind that animal technicians and mice help produce alongside one another. Knowing in the animal facility is subjugated, as the relationally produced mind of mouse and technician and cage is repressed so that its world-making elements can be denied. It did not seem to me during my research that animal technicians experienced alienation because they feared losing their job. None of the technicians I met articulated anything like job insecurity. Rather, the employer at the animal facility expressed concern about maintaining their animal technician staff; turnover was the problem and becoming a good place to work was the solution (see Friese 2019). Rather, alienation occurs when the relational knowledge produced in the animal facility is not recognized or considered significant because the only knowledge that matters is scientific knowledge.[11] Microsociology can help us be attentive to the time that goes into invisible forms of knowledge, and can facilitate an appreciation of these forms of knowledge that are at systematic risk. Care as knowledge requires habituation to the indistinguishable so that shock or surprise can make that embodied knowledge realizable. The indistinguishable is required for people like Martine to put into words or symbols the indices of interaction they accumulate over time in conducting their knowledge as care and understanding what makes their work meaningful in the process.

The ethics of animal care scholarship critiques how animals are treated in society (as food, pets, experimental subjects, and laborers), in line with animal rights advocacy but through a focus on relations instead of universal precepts (Narver 2007). Scholars of the ethic of animal care are less informed by the social science research methods that Gilligan used and are more likely to draw on the normative values of relational thinking that she advocates (e.g., Slicer 2007; Gruen [2004]2007; Donovan 2006; C. J. Adams [2006]2007). For feminist philosophers and activists, this scholarship emphasizes the importance of sympathy in creating better relations with animals. It seeks to recuperate the emotive at a time when affect and feeling have been disparaged by animal rights advocates such as Peter Singer.

Drawing on this strand of ethics of animal care, I contend that care can lead to ends in animal use, such as seen in the case of Elspeth. However, "care" and "animal use" are not irreconcilable. Animal technicians like Martine and Janet care a lot about animals, and they still "use" animals in a science that inflicts pain and death.

Humanitarianism through the Prism

What are the implications of these hierarchies of knowledge for humanitarianism?

Humanitarianism tends to move from crisis to crisis (Rieff 2002). Short-term action does not allow for the creation of a relational mind, as described in this chapter, and thus the care/ knowledge this implies. Those with this knowledge are regularly local workers contracted by international humanitarian organizations. Local workers are often the more marginal decision-makers (Fassin 2005). Not unlike the animal technician, their knowledge is the most likely to be rendered invisible and marginal as "local" or "intimate" knowledge (Raffles 2002). If we can ask what the life sciences would look like if the relational mind produced in the animal facility were central to epistemologies rather than rendered invisible, we can also ask what humanitarianism would look like if the work of country-level workers were at the center, rather than the periphery, of decision-making.

Without emplaced people working day in and day out in caring relations that are embodied and practiced, humanitarianism of any form—human and more-than-human—will wither away. It is important to highlight the kinds of caring practices that often go unseen and to value these knowledge practices as knowledge. The structural elements that make this knowledge invisible need to be understood as not solely political and economic, but also related to epistemic hierarchies and the *longue durée* of how knowledge is valued in professions and fields. In this context, hierarchy is a problem for both humanitarianism and more-than-humanitarianism. The benefit of more-than-human humanitarianism is that it can critique hierarchy as something that produces inequities, without forgoing the importance of differences in the process. We can see the importance of allowing different ways of knowing to be practiced alongside one another, so long as both are valued.

3

Killing

The killability of animals is where humanitarianism and more-than-human humanitarianism have radically different starting points. As Astrid Schrader (2017, 50) states: "Our relationship to death is perhaps one of the most enduring characteristics used to differentiate humanity from animality." The social fact that humans are not killable is a historic achievement rooted in Kantian ethics; it is an aspiration that has unfortunately not been fully realized in practice to date.[1] Nonetheless, in their different relations to killability, the human of humanitarianism and the animal of more-than-human humanitarianism are decidedly different. One of the strongest arguments for human exceptionalism in humanitarianism and human rights is that it provides a logic through which humans cannot be rendered killable. Humans are to be saved from death, and animals are to be killed in the process. Laboratory animals are, in contrast, to be saved from suffering, and often this is done by killing an animal who suffers (Bubeck 2023; Roe and Greenhough 2023).

What distinguishes animals from humans in more-than-human humanitarianism is, therefore, not only killability per se but also a very different relationship to saviorism. Humanitarianism is focused on saving lives and is thus part of a biopolitical project rooted in making people live in ways that are healthy and, by extension, productive. Foucault ([1976]2020) showed how power shifted in Europe from the sovereign exerting power by making some people die and letting other people live to an alternative set of power relations focused on productively making people live, but wherein some are allowed to die. This does not mean that making some die has disappeared, however. It is well established that biopolitics relies upon forms of "necropolitics" (Mbembe 2019).[2] Laboratory animals are a site of necropolitics that facilitates the biopolitical project of saving human lives through medicine and health care. More-than-human humanitarianism seeks to respond with care to the social fact of killing animals. Saviorism is not to be an option here; ameliorating suffering is.

Death is a pervasive element in biomedical research involving laboratory animals. All the mice I witnessed would eventually be killed for research unless they died beforehand. There has been significant discussion of adding a fourth R to the 3Rs, where the mandate to rehabilitate or rehome laboratory animals would be added to the mandate to replace, reduce, and refine laboratory animals (Pereira and Tettamani 2005). However, there is little interest in rehabilitating mice in the way that laboratory primates are rehabilitated to a sanctuary (Sharp 2019), or laboratory cats (Friese 2013) and dogs (Skidmore 2024) can be adopted into homes following laboratory experimentation.[3] When I asked a scientist about this, she responded: How would you rehabilitate the over one million mice used in experimental research in one year in the UK alone?[4] The sheer number of laboratory mice is seen as a barrier to generalizing rehabilitation in the ethical and moral landscape of animal experimentation. Mice are, after all, also considered pest and prey; they are "killable" (Haraway 2008) as a matter of their species being in a way that primates, cats, and dogs are not.[5]

How humans kill animals as part of an experimental process has implications for both the people involved (e.g., Tallberg and Jordan 2021 (Online); Birke, Arluke, and Michael 2007; Sharp 2019) and the knowledge that is produced (Lynch 1989; Svendsen and Koch 2013). This has been the focus of much empirical research to date. This chapter shifts the focus to the fact of maternal infanticide among laboratory mice; by probing the "problem" of mice killing other mice, this chapter explores who can or must kill and how, and who can't kill.

First, however, this chapter intervenes in a set of debates within animal studies based on J. M. Coetzee's novel *Disgrace* ([1999]2010). Rather than using fiction to amplify ethnographic moments and resonate the cultural logics and affects of more-than-human humanitarianism, this chapter and the next instead use ethnographic vignettes to intervene in a set of debates within animal studies that has already used fiction. Coetzee's novel has indexed debates between animal rights and what is at times referred to as "posthuman ethics." The debates reverberate between those who use their research to critique and resist the killability of animals (i.e., animal rights) and those who use their research to describe and make visible the scale of animal death that human life relies upon (i.e., posthuman ethics). As a scholar, I don't use the term "posthuman" but

I would fall into this latter category. Nonetheless, I greatly respect and engage with scholarship in the first category. Therefore, my goal here is not to pit two different social science approaches against one another, but rather to open up a new set of questions. By holding these perspectives together, simultaneously, I investigate "killing relations" rather than focus on what deems the animal "killable." This intervention captures the complex interplay between those relations that are rooted in killing and the potential violence involved in "killing" or ending such relations—whose institutionalization we would like to see come to an end.

Killing and Killability

The human–animal boundary is most clearly delineated through practices related to death, and more specifically practices related to killing (e.g., Haraway 2008). Henry Buller (in Schrader and Johnson 2017, 11) summarizes: "Killing is the original ontological act not just because it renders only animals uniquely bare-of-life and thus killable but also because it makes humans uniquely response-able in how we kill nonhumans." The social fact that nonhuman animals are killable, while human animals are not, represents a stark difference and divide between more-than-human humanitarianism and humanitarianism. The killability of animals is what gives more-than-human humanitarianism its impulse to take responsibility for how nonhuman animals are killed. In accepting that—under current institutional arrangements—laboratory animals will be killed, the questions shift to ensuring that their lives are worth living, their suffering is minimized, and that these animals will hopefully die good deaths.[6]

Taking responsibility for the lives of laboratory animals is in line with Donna Haraway's (2008, 1991) argument for feminist politics that tries to "stay with the trouble" of power relations between humans and nonhuman animals. Haraway states:

> Staying with the complexities does not mean not acting, not doing research, not engaging in some, indeed many, unequal instrumental relationships; it does mean learning to live and think in practical opening to shared pain and mortality and learning what that living and thinking teach. The sense of cosmopolitics I draw from is Isabelle Stengers's. . . .

> Stengers insists we cannot denounce the world in the name of an ideal world.... Stengers's cosmopolitical proposal, in the spirit of feminist communitarian anarchism and the idiom of Whitehead's philosophy, is that decisions must take place somehow in the presence of those who will bear their consequences. Making that "somehow" concrete is the work of practicing artful combinations. (Haraway 2008, 83)

We cannot escape histories of violence and oppression, and moreover, any attempt to do so risks erasing rather than ameliorating present and past, large and small forms of violence. For this reason, Haraway does not start with an ideological opposition to killing animals, as animal rights activists often do. Rather than trying to change the world by separating oneself from it, Haraway starts with complexity by sharing suffering with other people and species to forge better ways of becoming together. Haraway's figuration of staying with the trouble in laboratory science is the scientist who shares the suffering inflicted upon the experimental animals.

This chapter, therefore, joins a significant body of literature that has sought to lay bare and highlight the magnitude of animal death that human health relies upon. Lynda Birke, Arnold Arluke, and Mike Michael's (2007) book *The Sacrifice* showed how death is normalized as part of laboratory work. Tora Holmberg (2011) as well as Emma Roe and Beth Greenhough (2023) have explored how good deaths for laboratory animals are achieved by not only reducing pain but also by reducing the animal's anxiety and distress. Roe and Greenhough (2023) as well as Lesley Sharp (2019) show how distressing killing animals is for both technicians and veterinarians as well as scientists. Supporting one another when killing becomes "impossible" is crucial to how people care for one another in animal facilities (Roe and Greenhough 2023, 61). And as Sharp (2019) shows, many people will drop out of animal research as a result of the impossibility of facing an animal—day in and day out—whom one will kill.

This body of research sits in tension with activist-oriented scholarship that critiques in order to end the usability and killability of animals (Giraud and Hollin 2016; Giraud 2019, 2024; Gruen [2004]2007; Donovan and Adams 2007; Adams [2006]2007; Slicer 2007). The feminist ethic of animal care starts with a different set of questions: to sympathetically ask what the animal as a communicative other says. It argues that all animals will say that they do not want to be killed or eaten, even if done so humanely (Donovan 2006; Donovan and Adams 2007, 13). On this basis, this scholarship tends to start with opposition to veterinary work as it is used with livestock (Donovan and Adams 2007; Gruen [2004]2007) and laboratory animal science (Slicer 2007). Arguing against animal experimentation within an edited volume on the feminist ethic of animal care, the American philosopher Deborah Slicer states:

> I am not saying that everyone who cares about laboratory animals will condemn experimentation. I am saying that we will at least cease to condone the practice so cavalierly. We will find that there are certain elements of moral tragedy in having to make some choices despite the daunting complexity of these situations, despite having few, if any, principles or precedents to guide us, despite having little or no assurance that we have chosen rightly. And regardless of how we choose, we may have to live with, as some have recently put it, irresolute, nagging "moral remainders." (Slicer 2007, 120)

Slicer notes complexity rather than addressing how it is managed; this positionality doesn't seek to give space to how people (including scientists) engage with ethical dilemmas in such contexts. However, while doing this research, I noticed that everyone I encountered dwelled in this space of complexity, conflicting moral demands, and moral tragedy. The question of how people occupy this space of complexity when living with, caring for, and even killing animals is addressed by becoming entangled with the people who do this work (see also Atwood-Harvey 2005).

The research that makes this book possible is thus also embroiled with the more-than-human humanitarianism that I seek to name and describe. To explore this tension, I engage closely with animal rights activist Kari Weil's (2006) critique of Coetzee's *Disgrace* and her argument that making the killability of animals visible is a flawed project, as it fails to critique the fact that animals have been deemed killable.[7] I counter that Weil's theory of change fails to address that there is violence present not only in relations that are rooted in killing, but also in the processes of killing or ending those relations. Situating these relations accounts for this interplay across different contexts and scales of relating.

Killing Relations

Kari Weil (2006) has used Coetzee's *Disgrace* to develop a critique of the killability of animals and to trouble the "everyday moralities" (Sharp 2019) that such killings require. In critiquing the novel, Weil also critiques research that I engaged in, which is rooted in exposing and pondering the vastness of animal death that is otherwise left invisible and unremarkable. Her critique is important. Weil suggests that the protagonist of *Disgrace*, David Lurie, experiences a personal transformation through the act of killing dogs humanely. He kills dogs who are simply too reproductively prolific to live well in the South African postapartheid countryside. "The animals they care for at the clinic are mainly dogs.... The dogs that are brought in suffer from distempers, from broken limbs, from infected bites, from mange, from neglect, benign or malign, from old age, from malnutrition, from intestinal parasites, but most of all from their own fertility. There are simply too many of them" (Coetzee [1999]2010, 142). Weil argues that Coetzee makes this extent of animal death visible but fails to critique it.

> It is important, indeed critical, to be open to affect, to what we do not know; it is what calls us to ethics. But affect responds to and calls upon potentially unethical drives and passions. An animal-other may call us to our responsibility, and we may interpret that responsibility as "thou shall kill" or "thou shall not kill." We cannot know for sure which is right; all we can do is to attempt to listen and respond through an act of empathy that may require becoming someone or something we have never been and imagining a response that is other than we have known. (Weil 2006, 96)

Weil calls for animal studies that does not only expose the dilemmas of animal killing through relational approaches to ethics, but also that challenges animal killing as wrong.

I seek to respond to Weil's critique of killability by parsing different killing relations. Focusing on killing relations does not address killability, in the singularity of one act, without reflecting on the scale of killing relations it is entangled with. I therefore put David Lurie's killing practices within the larger context of the postapartheid moment in which the novel is set.

Professor David Lurie is a disgraced academic, an old-fashioned philanderer caught out following a highly coercive sexual affair with a student that ends in rape. David Lurie will not repent but accepts his punishment, leaving the university and Cape Town to visit his daughter Lucy in the countryside on her farm. Lucy is selling part of her farm to Petrus—a black African man. One day when Petrus is away, three men arrive at the farm asking for a phone but beat and burn David, rape Lucy, and kill all of Lucy's dogs. David makes his beating public by going to the police. He wants Lucy to make her rape public as well, but Lucy is adamant that, in the current place and time, her rape must be a private matter. Lucy tries to explain to David that she is being subjugated and will not run from the punishment of history.

Lucy may be sorry for the history of apartheid, and David may be sorry for his own sexism and misogyny, but being sorry is not the question of this novel. David goes to the father of the student he raped to give him an apology. Her father tells David: "You are sorry.... But I say to myself, we are all sorry when we are found out. Then we are very sorry. The question is not, are we sorry? The question is, what lesson have we learned? The question is, what are we going to do now that we are sorry?" (Coetzee [1999]2010, 171–72). David and Lucy's relationship crumbles around this question. Lucy is pregnant following the rape, and one of the rapists is living on the farm with Petrus. Lucy will not leave, nor will she abort the child. She instead considers becoming a third wife to Petrus, keeping her home and giving Petrus the farm in order to gain protection in the new landscape of postapartheid South Africa. David cannot fathom his daughter's response, and he wants Lucy to leave and escape to the Netherlands, where her mother lives. At this impasse, David is left to live nearby in Lucy's shadow, to keep an eye on her as he helps Bev Shaw, a veterinarian, kill surplus dogs.

No one in the novel can escape how their personhood is lost as they become an index of apartheid: David is a white man, Lucy is a white woman, Petrus is a black man, and dogs—long used by and for the police—are nonhuman agents of the apartheid state. Calina Ciobanu (2012, 679) summarizes: "The singularity of the individual is thus obliterated as each being is converted into the emblem of a type and punished not for what it has actually done, but for what it represents." How do we engage with Weil's critique of the killability of dogs when put back within the postapartheid frame of the novel itself—where the killability of humans and animals is also about the vexed question of how to kill the relations of apartheid in everyday practices?

This point is crucial. The simple fact is that if "laboratory animals" were "killed" as a relation, a great many individual animals would need to be killed as a result. *Disgrace* does not simply describe animal killing without making killability problematic. Rather, the novel says that killing or ending relations, especially those relations that *should* be killed such as sexual violence and apartheid, is nonetheless likely to be violent. Killing one relation can impede killing another; sexual violence may be rooted out in the university but persists on the farm; racism may be rooted out on the farm, but dogs are still deemed killable. The novel asks if we, as readers, can find it in ourselves to respond with some compassion to such a dislikable character and perpetrator as David Lurie. This compassion is required if killing those relations that *should* be killed might be done with slightly less violence to the "singularity of individuals"—human and animal individuals alike.

In contrast to Weil, I would argue that David Lurie's transformation is not in killing dogs per se. Rather, I see his transformation as occurring in mourning those dead dogs—dogs he cannot save. Lurie finds himself responsible for bringing the dead dog bodies to the incinerator at the hospital. He begins to load the dog bodies onto the conveyor belt himself, as he wants to ensure their dead bodies are treated with respect. He cannot bear seeing the dead dog bodies beaten with shovels to fit the conveyor belt with more ease. Lurie thinks:

> Why has he taken on this job? To lighten the burden on Bev Shaw? For that it would be enough to drop off the bags at the dump and drive away. For the sake of the dogs? But the dogs are dead; and what do dogs know of honour and dishonour anyway? For himself, then. For his *idea of the world*, a world in which men do not use shovels to beat corpses into a more convenient shape for processing. The dogs are brought to the clinic because they are unwanted: *because we are too menny.* That is where he enters their lives. He may not be their saviour, the one for whom they are not too many, but he is prepared to take care of them once they are unable, utterly unable, to take care of themselves, once even Bev Shaw has washed her hands of them. . . . Curious that a man as selfish as he should be offering himself to the service

of dead dogs. There must be other, more *productive* ways of giving oneself to the world, or to an idea of the world. One could for instance work longer hours at the clinic. One could try to persuade the children at the dump not to fill their bodies with poisons. Even sitting down more purposefully with the Byron libretto might, at a pinch, be construed as a service to mankind. But there are other people to do these things—the animal welfare thing, the social rehabilitation thing, even the Byron thing. He saves the honour of corpses because there is no one else stupid enough to do it. That is what he is becoming: stupid, daft, wrongheaded. (Coetzee [1999]2010, 146, emphasis in original and deliberate misspelling of "too many")[8]

David Lurie is doing dirty work for the first time in his life, work that no one else will do. He is doing the work of social reproduction, making "an idea of the world" that is decidedly not "productive" but trying to enact a better idea of the world within existing constraints. It feels like stupid work to him, given David Lurie's place and time in history, but this work is demanded of him—mourning the killability of dogs.

In erasing the postapartheid context of the novel and the work of mourning, Weil suggests that the killability of animals is made too acceptable in the novel. By extension, this critique applies to the corner of animal studies that I inhabit, wherein killing is deemed to be too readily accepted. However, Coetzee's novel is not about killing dogs; it is about killing relations—killing the relations of apartheid that dogs were entangled in, as part of the policing that violently sustained apartheid. The question Coetzee asks is: How might we address the disgraces of history without turning people and dogs into indices of those disgraces? The challenge put before us by Coetzee's novel is the central challenge of humanitarianism and more-than-human humanitarianism alike, especially in these troubled times we are living through today.

I will now turn to a vignette that shifts the focus from the killability of animals to killing relations in their multiple forms. Infanticide is a relation that is rooted in killing. It is a relation that most want to end or kill because it is an act that can only be viewed as denoting a significant problem.

Vignette

I am spending the day following a postdoctoral researcher named Alana, who is conducting cancer research. Alana is comparing 12-week-old healthy mice to 12-week-old mice with the deletion PTEN. This gene codes for the protein phosphatase and tensin homolog, a tumor suppressor commonly lost in human cancer. Alana also compares 10-month-old healthy mice to 10-month-old mice with the deletion PTEN. The 10-month-old mice with the deletion PTEN have all had prostate tumors she is interested in understanding. The goal of her research is to make tags for the proteins involved in tumor development, particularly protein P13 kinase, three families of P10, and three families of P8.

Alana is at this point five years into this research project and is nearing the end of her funded research. She tells me that the first four years had been spent getting all the mouse lines, including aging to 10 months the healthy mice and—more problematically—the genetically modified mice with the PTEN deletion. Both the breeding and the aging occurred in the transgenics unit of the animal facility. Because of the nature of breeding, Alana told me that only one out of every eight mice born had the correct PTEN deletion. In other words, Alana could only expect one mouse in every litter born for her research to have the correct PTEN deletion and thus be usable. In addition, and to make matters worse, mortality rates were high among the mice that did have the PTEN deletion, and this began from birth and continued across the 10-month life course. Getting enough mice to live to 10 months with the PTEN deletion was a significant hurdle for her research and took most of her energy for the first four years of her funding.

In the middle of the day, Alana and I take a break to meet the rest of the members of her laboratory for birthday cake and coffee. While chatting over cake, another scientist, Daniel, asks me about my research. I tell him that I am interested in how care figures in the work of life scientists because this work is often an important but unrecognized aspect of scientific knowledge production. Daniel responds that he agreed with me: animal care is a real and unspoken problem for biomedical research, and he thinks it represents something like an "elephant in the room."

Daniel goes on to define the problem of animal care in science in very specific terms to me: as the preweaning death of infant mice inflicted by the mother. He tells me that the problem is that everyone thinks there is a litter of mice in the animal facility, and they leave the mouse in peace to nurse until it is time to wean the pups. No one can see what is happening in the cage during this time because the infant mice are covered with the cage bedding (essentially shredded paper), which the mother has turned into a nest for her pups. But when it comes time to wean the infants, the cage is empty except for the mother. The assumption is that the mother must have eaten her offspring.[9]

Daniel further tells me that he thinks the problem of mouse infanticide is located specifically within the animal-facility environment and is linked with mouse behavior that develops within this environment. He notes that the foster mothers in the animal facility are often of a different strain of mouse than the infants, so this behavior may be strain specific and linked to genetic inheritance and thus, the creation of transgenic mice. Before getting up to go back to work, he comments that I am shadowing the right postdoc. The PTEN cancer research has undergone serious delays and difficulties because of the preweaning infant mortality of mice.

Alana's difficulties were, therefore, not limited to the nature of her specific research and the biology of cancer. Her difficulties were, according to Daniel, amplified by a more general problem in animal facilities and possibly the sociobiology of transgenic mice, the "biosociality" (Rabinow 1996) of transgenic mice, or both.

* * *

The problem of newborn mice dying after birth and before weaning was also felt within the animal facility. I am spending the day with Vincent, one of the two full-time veterinarians at the Institute. Vincent is taking me to the transgenics unit to show me the facilities and think about how he will set up a new research project to better understand preweaning infant mortality in mice. Like Daniel, Vincent tells me that infant death among mice is a problem for animal facilities. Unlike Daniel, Vincent does not think the problem is attributable self-evidently to the animal-facility environment. For Vincent, the problem is that they simply do not know what is happening in the cages between the mother and the infants during this time. To address this problem, Vincent is setting up cameras to record this time period in the mouse life cycle within the animal facility. He wants to better understand what happens between mother and infant mice during nursing.

Vincent and I make our way to the transgenic unit. The manager of this unit, Rose, is called to meet me in the locker room, and she explains that I need to take off my clothes and put them in a locker before showering with the provided soap, shampoo, and conditioner. A towel, scrubs, socks, and shoes are available on the other side of the shower. I dress, put my hair up with a provided clip, and cover my wet hair with a net. Leaving the women's changing room, Rose and I enter the transgenics unit of the animal facility and join Vincent.

Rose leads us to a busy room where many mice are kept and much of the breeding work takes place. The room is lined with cages, and technicians are busy at the hoods in the middle of the room. They are busy changing the bedding of the mouse cages and checking the health and well-being of the mice. Vincent takes this opportunity to look over the cages with Rose, discussing any particular issues that have recently arisen. Along the way, he points out the cages marked "FOSTER MOTHER" on the card at the front of the cage. I see here the foster mothers caring for the infant mice of those biological mothers who had a history of eating their offspring.

Vincent leads me to another, far quieter room where he plans to set up his experiment. There are two empty racks for mouse cages at the front of this room, which he is hoping to use. He looks at the racks to see where a video camera can be set up to view the brood nest. After deciding where he wants to place the cameras, Vincent notes that the next step is to set up a test to ensure that he can see within the brood nest from that perspective. Seeing the mother with her pups will help him to better understand why the infants are dying.

Infanticide

In our discussions, Vincent regularly used ethological knowledge to question and critically interrogate what is considered a "life worth living" for laboratory mice. Inspired by our conversation, I decided to do a nonexpert review of the state of ethological knowledge on infanticide in mice. It appears that, at least during the final quarter of the 20th century, ethologists, for the most part, believed that male mice would kill newborn pups to whom they were not related but that female mice would not. This belief was supported by the experience of animal technicians working with laboratory mice, as female laboratory mice were then only very rarely reported to exhibit infanticide (McCarthy and Vom Saal 1985). This belief starkly contrasts with reported experiences today (Weber, Olsson, and Algers 2007).

Later in the 20th century, ethologists began to find that wild caught female mice would exhibit infanticide at certain points within their reproductive life cycle (McCarthy and Vom Saal 1985). Infanticide was seen in the context of certain kinship arrangements, specifically those involving communal nesting, wherein female mice who were not the mother were the primary predators for newborn pups (Vom Saal et al. 1995). These female mice could, in other instances, also be an important source of support for newborn pups, at times increasing pup survival rates (Weber, Olsson, and Algers 2007). In this context of conflicting evidence, the idea was that female infanticide was seen in wild caught mice, but not among mothers and only under certain conditions. The presumption here is that female mice will only kill newborn pups whom they are not the biological mother of, and only while under stressful conditions.

Infanticide behaviors among maternal mice are thus, according to ethological knowledge, something of a 21st-century problem. This behavior is currently thought to result from stress in the mother's environment. But that stress could come from two very different sources. It could be that something in the environmental conditions is creating maternal stress, which results in the mother actively killing her offspring. However, there is also evidence that some specific laboratory strains of mice, particularly genetically modified strains, have poor maternal behavior (Weber, Olsson, and Algers 2007). So, it could be that some strains of transgenic mice are unable to care for their infant pups and the pups then die from either hypothermia or starvation. In this case, the mother eats her dead pups rather than actively killing them.

These ideas are circulating in laboratories, among animal technicians and veterinarians beyond the Institute where I conducted observations. Lesley Sharp (2019, 145), for example, quotes an experienced animal technician who articulates this ethological knowledge in terms of her everyday work: "Someone will come in and say 'something's wrong with my mice—they keep dying' and I ask them, 'Well, what are you doing when a dam has new pups? Do you keep opening the drawer and checking up on her? Because that will stress her out to no end, and her natural response is to protect herself when she thinks she's in danger, and the first thing she'll do is eat her pups. If you don't need them, why do you keep looking in on them? Leave them alone and she'll take good care of them.'" Vincent is thus trying to see the mother with her pups, without creating stress, to better understand the nature of the problem that is leading to infanticide and thus ameliorate the conditions giving rise to this behavior. He needs to know if there is a breeding problem or if some practice within the animal facility is creating stress for the mother.

For many people infanticide is an abhorred practice, one that has a long history of being used to legitimize further forms of violence against the perceived perpetrator.[10] Lesley Sharp (2019, 141) quotes a doctoral student in genetics who stated: "I detest mice. They are cannibalistic and will quickly resort to eating their young. This is why I work with rabbits." Here, we see the indictment that there is something about mice that makes them commit infanticide. Figuring mice this way allows people to detest them as a species, making them "killable." However, there is also the lurking concern, which Daniel articulated, that something about the animal facility compels a mouse to kill her pups. Is it the case that the mother decides that her pups' lives are not worth living in the laboratory? If so, this has far more profound implications for the animal facility and scientific research. In either case, infanticide is deemed to be something that should not happen. There is something wrong with either the mouse strain or the environment if infanticide is occurring. Infanticide as a killing relation must be ended—according to either of these sets of ideas.

On multiple occasions, when I presented this vignette at academic conferences and workshops, people made an intertextual reference to Toni Morrison's novel *Beloved* (1991). I believe this novel allowed people to reckon with the fear that arises when one asks how wretched a life must be for a mother to kill her newborn pups. *Beloved* is about a former slave and mother who is haunted by having killed her newborn daughter in order to save her from the horrors of slavery. From a humanist perspective, this intertextual reference that connects laboratory mice with human slavery is entirely inappropriate. While some have made the argument that domesticated animals are the equivalent of human slaves (Spiegel 1996), I would argue against the logic of commensurability that this claim is rooted within. Rather, I suggest that chattel slavery in the Americas is a narrative of violence that creates a "site of recognition and reckoning" (Bennett 2020, 11) for infanticide in laboratory mice. Evoking this novel was a way for people to imagine what it might be like to be a laboratory mouse—breeding offspring to be used and ultimately killed for the betterment of another species. Connecting with *Beloved*, I believe, has been an act of imagination by my colleagues that allows them to ask if these mice are killing their offspring to save their newborns from becoming laboratory animals. This emotive questioning is useful to register discomfort; it is a compassionate site of anthropomorphism.

What ethological knowledge seeks to do is to temper the affective response that the reference to *Beloved* gives rise to, by asking questions about the specific nature of the problem of mouse infanticide that is at hand. There are two very different welfare consequences of maternal infanticide in mice, depending on whether the problem is genetic or environmental. Foster mothers are an appropriate response if the problem is genetic/strain-based. Ameliorating the environment of the animal facility so that it is not stressful to a brooding mouse is an appropriate response if the problem is behavioral. Focusing on ethological knowledge, therefore, means sidestepping the existential questions about whether a laboratory mouse considers her life worth living.

More-than-Human Humanitarianism and Killing

The ethical mandate to "replace" animals from research articulates the goal of putting an end to using and killing animals as part of science. Replacement means killing the "laboratory animal" as a human–animal relation. However, science and technology studies (STS) is rooted in attempts to be symmetrical in our analyses (Bloor 1991). The figure of the mouse who kills her own offspring sits uncomfortably with this goal of replacement. She evokes the feeling that the mouse, too, may be trying to kill the laboratory animal as a human–animal relation by killing her own relations. Therefore, if we view the actions of the mouse killing her relations as in need of problematizing, we also need to at least question the social killing of the "laboratory animal" as a relation.

What if we no longer use animals in research? This would undoubtedly be a good thing. But how do we get there? Many animals will die, and many animal strains will go extinct as, crucially, transgenic strains of mice cannot live outside of science. Knowledge about how to take care of these animals could disappear. There could be negative consequences for the state of biomedical knowledge, making the precarious health of some humans that much more precarious. It is important to stress that none of these are reasons *not* to replace animals from research. However, it is important to problematize and highlight the assumptions that are made: that extinction doesn't matter if the strain is human-produced; a life in science is not a life worth living; the embodied knowledge of animal technicians does not need to be preserved. Certain inequalities are reproduced in the very valiant process of replacing animals in the production of scientific knowledge.

Being able to witness the killing of animals without displaying physical and emotional distress was a requirement for doing this research as a social scientist, and I did feel that I needed to prove myself in this respect. Just before the first mouse experiment that I was ever to watch, Elspeth turned to me and asked, "You're not going to faint on me, are you? It's okay if you do—sometimes a new PhD student will. But I can't do anything to help you because I'll be with the mice." I responded that I did not think I would faint, and indeed I did not. But I was happy to have had the warning; I willed myself not to faint. In doing this research, which has in part sought to understand how people can both care about animals and kill those animals, I have been interpolated into a side within an academic debate as I have had to become complicit in killing animals in order to understand this practice.

However, to say that more-than-human humanitarianism is complicit in the killability of animals is only a partial truth in this context. More-than-human humanitarianism *is* "partially connected" to animal rights by supporting replacing animals in scientific research. What more-than-human humanitarianism is centrally concerned with is the more vexed question of: How? To this I would add: What are the potential costs for differentially situated actors? Does a transgenic mouse have an interest in a future life, even if that life can only be within science? Maybe these animals' lives are stage-managed in relationship to their death, and for the purpose of humans. However, as Henry Buller (2013) pointed out, the only means of escape for laboratory animals is either to die or to be unborn. How might we kill a relation like the "laboratory animal" with some kindness, recognizing that the work of animal technicians and their knowledge disappears with the disappearance of "laboratory animals" as a relation?

It is important to emphasize here that the cage has been a key site through which the well-being of laboratory animals was sought, something that Robert G. W. Kirk (2016) has shown was largely driven by the belief that welfare is the absence of infection. Kirk contends that cage design thus enacted knowledge and beliefs about the conditions that produce pain and promote animal well-being; the cage embodies very particular moral economies (see also Druglitrø 2016). The cage is thus certainly a site of human domination and potential suffering (Giraud and Hollin 2016). Still, this totalizing categorization risks erasing how cages are also interfaces in human–animal interactions that embody versions of care and management (Bjorkdahl and Druglitrø 2016, 7). Kirk (2016) emphasizes that the laboratory mouse or rat was increasingly domesticated as distinct from its wild counterparts during the mid-20th century. These animals were no longer collected from the wild but had been bred for several generations for the purpose of science (see also Rader 2004). In this context, the cage became the milieu for these new kinds of domesticated species and the only environment where these strains could live. Infanticide means that something is wrong with the mice and their cage; mice cannot kill other mice in the context of laboratory animal science.

Humanitarianism through the Prism

The social fact that animals are killable raises both connections and disjunctures between humanitarianism and more-than-human humanitarianism. The human subject of humanitarianism is deemed not killable by the right of their humanness. Saving the human subject from being killed, and from being deemed killable, is the central goal of humanitarianism. Meanwhile, the nonhuman animal subject of more-than-human humanitarianism is deemed killable by the right of their nonhuman status. Tarquin Holmes and I (Holmes and Friese 2020) explored the crucial role of anesthesia in the 1875 Royal Commission on Vivisection precisely because it helped institute animal pain as the key problem of animal research, as opposed to animal death. In being made killable, the nonhuman animal should not suffer. In being made to live, the human, by extension, can and will suffer (see Stevenson 2014).

Killability not only distinguishes humanitarianism from more-than-human humanitarianism but is at the same time also another site where this binary is "one approach [that] is only ever partial" (Strathern 2004, xiv). In both humanitarianism and more-than-human humanitarianism, specific modalities of killing are justified through the law by delimiting who can kill another and how.[11] Veterinarians can kill animals. The armies of nation-states can kill people. Moreover, there is tension and conflict within humanitarianism over the killability of humans, such as in "humanitarian wars"—wherein civilians are strangely made killable in the name of their human rights, as seen, for example, in the US discourse about the war in Iraq (Cubukcu 2017; Rieff 2002).

Similarly, while laboratory animals may be killable, they cannot be killed by anyone or in any way. The Institute and every scientist working within the Institute needs to have a license from the Home Office that permits them to use animals in research. That license covers only specific ways of making animals live and die as part of science. Killing cannot be done or decided upon by anyone. A mouse is not allowed to kill her pups in the animal facility as this presents both a welfare and a science problem. Documentary moralities (Druglitrø and Asdal 2024) are part of a regulative structure that makes laboratory animals killable in specific ways according to the law, and accompany the everyday moralities involved in killing animals (Sharp 2019).

The law also makes humans killable by certain actors and under certain conditions. War is just one example of this. In this sense, what humanitarianism and more-than-human humanitarianism share is an ethos to respond to the fallout that arises when certain lives—human or nonhuman animals—are made killable under the law.

We can learn from more-than-human humanitarianism how ethical action is forged when saviorism is not an option. Saving lives may be the rationale of humanitarianism, and it may be its shining glory. However laudable saving life is, saviorism has also been shown to reproduce inequality. Second-wave feminism suffered from racism that underlay its approaches to saving women and girls from patriarchy. Humanitarianism equally suffers from racism and coloniality that underlies its approach to saving lives (see, e.g., Stevenson 2014). Becoming a humanitarian—a savior of life—may be seen as a way of creating an ethical self that one can be proud of (Givoni 2016). More-than-human humanitarianism is not rooted in this type of saviorism. David Lurie is not transformed by saving dogs but rather by mourning dead dogs. More-than-human humanitarianism responds to the necropolitics that underlies any biopolitical regime, in this case, the biopolitical regime of medicine. It bears witness to the social fact of necropolitics when no one else will, and engages in mourning the multiple and interlocking violences that occur in the social reproduction of killing relations.

4

Sacrifice

In the United Kingdom, since the 19th century, Jeremy Bentham's (1789) question—"Can they suffer?"—has been the key ethical question regarding both how laboratory animals should live and how animal life should be brought to an end.[1] Killing is used to end an animal's life when it is believed that the extent of physical and psychic suffering already does or will—in the future—result in a life that is not worth living. Killing is seen as a tool of kindness in this context, when used by licensed practitioners such as veterinarians. Throughout my ethnographic fieldwork, I became habituated to this impetus to end suffering, which made the near-constant presence of death in animal facilities feel unremarkable at times.[2] Indeed, I was surprised by the extent to which I was able to conform to the normalization of killing animals. That said, there was a moment when I was watching mice die, and their deaths affected me in a way that made me sad and, in turn, deeply uncomfortable with my sadness. This chapter begins with this moment, critically interrogating how my emotional response to mice dying—both my normalization of killing mice and the interruption of this normalization through a moment of incredible sadness—is linked to the discourse of "sacrifice" as it operates in the life sciences.

In this chapter, I build upon the idea that "sacrifice" is a moral economy of laboratory science (Svendsen and Koch 2013) by suggesting it is also an "affective economy" (Ahmed 2004), one that circulates through laboratories and across bodies. As an affective economy, sacrifice is shaped by a feeling of purpose and hope in contributing to something larger than the self. In developing the concept of affective economies, Sara Ahmed pushes against the everyday understanding of emotions as property residing in individual subjects. Ahmed has shown that emotions circulate in economic ways that bind, thereby collectivizing emotions. I argue that sacrifice circulates in laboratory science in part to make the pervasiveness of death manageable; sacrifice makes the killing of laboratory animals possible at an emotional level.

To demonstrate the importance of sacrifice as an affective economy, this chapter considers what happens in its absence—when animal lives do not result in data. The extent and importance of sacrifice as an affective economy became clear to me by my own and other people's reactions to the death of mice who were scientific "misfits" (Star 1990; Bowker and Star 1999), in the sense that these mice would not contribute to the scientific infrastructure of timeless, universalizable data. The extent of death around me became almost unbearable, and sadness set in when I could not rely upon sacrifice as an affective economy. Sadness gets lodged in specific and individual bodies, and this lodging—this cutting off—is what Ahmed argues is the crucial moment that marks out vulnerability within an affective economy.

According to Ahmed, the vulnerabilities that arise from the lodging of emotions in individual rather than social bodies are generally linked to social structures of inequality. I, therefore, conclude by asking what the implications are of "sacrifice" being such an important affective economy for more-than-human humanitarianism. I do this by engaging with how sacrifice appears in the novel *Disgrace* (Coetzee [1999]2010). Similar to Johan van der Walt (2005), this chapter arrives at the bind that Derrida (2008) confronted regarding sacrifice: sacrifice exists, and sacrifice cannot seem to not exist, and yet sacrifice should not exist. Drawing attention to the magnitude of the sacrificial logic that undergirds both humanitarianism and more-than-humanitarianism is itself a step, albeit unsatisfying. Whenever one seeks to minimize the significance of the sacrificial status of another, human or otherwise, one should become aware that one is engaging in everyday practices of sacrifice that enact inequity.

Sacrifice in the Life Sciences

There is a body of literature that describes how "sacrifice" in animal experimentation legitimizes the death of laboratory animals because this death allows for the generation of scientific facts that will result in the betterment of human health (Haraway 1997; Lynch 1989; Birke, Arluke, and Michael 2007). Hope for and anticipation of medical breakthroughs is seen to justify the killing of animals in preclinical research (Holmberg 2008, 2011). Laboratory animals are thus not simply killed, but are instead seen as sacrificed for the greater good of human beings.

Moreover, ethnographic research within science and technology studies has shown that "sacrifice" does not simply work as discursive legitimization. Michael Lynch's (1989) canonical ethnographic research in a neuroscience laboratory has been crucial for understanding how sacrifice also operates in the everyday practices of bioscientific knowledge production. Scientific research involving animals requires a distinction and a relationship between what Lynch calls the "naturalistic" and the "analytic" animal. The naturalistic animal refers to the whole animal of the commonsense lifeworld; this is the animal that technicians and veterinarians encounter and are concerned with. The analytic animal, in contrast, is data in the form of a tissue sample, an electron micrograph, or a statistic based on the naturalistic animal's body. The analytic animal therefore requires validation through rigorous testing. Lynch contends that "sacrifice" is the pivotal moment in transitioning from a naturalistic to an analytic animal. Animals have to die in a specific and well-orchestrated way for an analytic animal to result. Sacrifice, therefore, occurs in the laboratory as a ritual—without being religious per se—wherein transcendence occurs through the practices that enable timeless data to be produced. Salvation is promised through treatments or cures for human illnesses and diseases.

Sacrifice may, therefore, be a uniquely scientific discourse. For example, Lesley Sharp (2019) found that most scientists she spoke with during her research in the United States and the United Kingdom used the terminology of "animal sacrifice" when describing their work. Scientists' use of this vernacular stood in stark contrast to veterinarians and technicians, who were more likely to use the tropes of "euthanizing," "putting down," or "killing" laboratory animals.

Vignette

Early on, when I had just started this research project, I visited an animal facility knowing that the mouse *model* a scientist would be working with had to be transferred to another institution. In other words, it was not the mice that needed to be moved but rather their specific genetic composition. Because of the biosecurity protocols of the receiving institution, the mouse model could not be transported in the form of living mice. The mice could carry a disease, which could then be spread to the receiving institution's mouse population. As a result, the model would have to be transported in the form of ovaries, where the ovary serves as a kind of biosecure container for the ova containing the genetic information of interest. The model would be reembodied in the new institution through in vitro fertilization as new and disease-free mice. What I was to watch that day was the recovery of ovaries from a large number of female mice.

Recovering the ovaries required killing the mice. It was not the case that the ovaries of the mice could be removed, and the mice could recover so that they might live to be in another experiment. There are strict rules governing the reuse of animals, as recovering from surgery is painful. While reusing animals may reduce the number of animals used in science, this would increase the amount of suffering that any one animal experiences. As a result, strict limits are placed on reusing animals, with the idea that an animal only has further interest in life if that life is not too painful.[3] Killing these mice was deemed the most humane way to retrieve their ovaries and protect other mice from potential disease.

The mice were being killed in the manner that everyone I spoke with told me was the most humane method, which is through dislocation of the neck. This may sound gruesome, but it is considered the kindest way of killing because it is fast from the mouse's perspective. There is little stress for the mouse before the procedure and little physical or emotional suffering resulting from the death process. Scientists would contrast the speed and lack of stress that dislocation of the neck ensures for the mouse to alternative methods, such as hypoxia. When hypoxia is used to kill mice, a gas such as CO_2 is added to a cage full of mice in order to remove oxygen. Scientists told me that while this method of killing a large number of mice is rather easy from the scientists' perspective, it produces high levels of anxiety and physical suffering on the part of the dying mice. Dislocation of the neck is kind because it moves the suffering from the mouse to the technician, as it can be emotionally difficult to kill mice in this way and it certainly takes more time from the technician's perspective. Killing a mouse by dislocation is, in many ways, a sign of sacrifice on the part of the technician, who takes up the stress of killing a mouse in close, physical proximity in order to give that mouse a good death.

My throat tightened, and tears began to well in my eyes as I watched these mice being killed and saw the growing number of dead mouse bodies pile up. I willed those tears to stop. I was thankful that we were in a dark room and that I was standing behind the scientist and technicians. I was unlikely to be seen crying. Not only would this have inhibited my ability to do this research, but I also worried that my crying could negatively affect the mice. I worried that my stress and sadness could become contagious and could be transferred to the mice just as these technicians were trying to take that stress away and give the mice a good death. I found myself incredibly sad and incredibly uncomfortable from my sadness. My internal distress was not the result of *how* the mice were being killed; they did not appear to suffer. I was instead affected by the assembly-line nature of the way these mice were being killed, the piling up of dead mouse bodies that then had their ovaries surgically removed.

I have wondered over the years why I became so sad at that moment and why I grieved so deeply for these female mice. I will admit that I have worried about my possible overidentification with the mice, as I was also in the processes of assisted reproduction at that time. But I don't think this is the case, or at least it was not the whole case. Rather, the story I have since told myself, and that I have been most convinced by, is this: I grieved for those female mice because they were being killed for their ovaries, for their heritable genetic material, which could be detached from their selves. The mice were being killed for their future reproductive potential; they were not sacrificed for their immediate scientific potential. It was only far later in the research process that I could articulate my emotional response in this way, and thus as bound up in the moral economy of "sacrifice" (Svendsen and Koch 2013) that is central to laboratory work.

I realized this several years later, during a day spent with Janet and the aged mice she cares for at the Institute. I asked Janet what happens if the techs and the vet feel a mouse will not make it to two years old and so cannot be used in the aging research. Janet replied that there was one mouse she thought would die before two years. She explained that sometimes the lab is told that they will just have to do something different because the mice will not live long enough for the experiment as originally planned; the lab then tries to do whatever they can. Janet then told me that it is hard to see a mouse who is starting to die and who won't quite make it to an experiment. "To have lived in these cages for two years or more, which must be really boring for the mice, and then to not quite make it to the study, it is really sad."

With Janet's conveyance of sadness, my previously isolated and isolating experience of sadness, in seeing mice killed for their ovaries, found a connection. On this day, my experience of sadness became decipherable when it was experienced as shared with Janet. I have come to understand this sadness as a response to the disappearance of the logic of sacrifice. I reconnected with sacrifice as an affective economy through its antithesis.

Sacrifice keeps sadness at bay most of the time, as people are getting on with the work of doing laboratory science. Sacrifice makes the extent to which (some) human health depends upon a mass of animal death entirely unremarkable (see also Svendsen 2022). But without this sense of purposefulness—without the sense of becoming data that can transcend life and death through universal knowledge, which in turn has the potential for salvation by creating more life for others—the vast extent of death bursts in. When death burst in, there was no other affective economy to rely upon.[4]

A Meaningful Way to Die

Mette Svendsen and her colleagues have analyzed sacrifice as a moral economy in laboratory biosciences that involve nonhuman animals as models (Svendsen et al. 2018; Svendsen 2022; Koch and Svendsen 2015; Dam and Svendsen 2018; Svendsen et al. 2017; Svendsen and Koch 2013; Dam, Sangild, and Svendsen 2020). They build on Lorraine Daston's (1995, 4) concept of moral economy in the history of science specifically, defined as a "balanced system of emotional forces, with equilibrium points and constraints." For Daston, these emotional forces denote a mental state that is collective rather than individual, embodying culturally mediated and historically specific values that inspire and shape scientific knowledge production.[5] Svendsen and Koch (2013) build upon this definition to show how entrenched the idea of sacrifice is for scientists and scientific work, such that sacrifice is not simply a legitimization strategy or ritualized practice but also a taken-for-granted idea that shapes everyday practices involved in doing research. Viewing sacrifice as a moral economy is a springboard for understanding the corresponding affective economy of sacrifice. In arguing that sacrifice is an affective economy, I explicitly build upon their empirical and theoretical research and offer an incremental development.

Svendsen and her colleagues, in particular Mie Dam (Dam and Svendsen 2018; Dam, Sangild, and Svendsen 2020, 2018), have conducted over a decade of ethnographic research on the use of pigs as models of human preterm infants in Danish biomedical science. They have shown that the pig is a rather unproblematic substitute for the human in Denmark, as pigs have long been central to the agribiopolitics of the Danish welfare state. Svendsen states (2022, 22): "In the beginning of the twentieth century, Danish pork production was elevated to a national project, as a way to produce wealth for the nation and let the population thrive on the pig as a national economic resource." Svendsen argues that the human exceptionalism of using animals as models for human health is entirely routine and unproblematic in the Danish context.[6]

Continuous with Lynch's conceptualization of sacrifice in laboratory science, Svendsen and her colleagues find that a pig must die a planned death at the time set by the experiment (Dam and Svendsen 2018; Svendsen et al. 2017; Svendsen and Koch 2013; Svendsen et al. 2018). However, Svendsen and her colleagues show that the moral ordering rooted in species difference and the prioritization of scientific values were not absolute. Scientists and animal technicians also cared for piglets as extensions of themselves (Svendsen 2022, 72) and as they would "their own child" (Svendsen 2022, 77). In this context, the question of when to euthanize piglets was not straightforwardly determined by scientific needs. Piglets who were suffering needed to be euthanized.

Lynch's conceptualization of sacrifice in the life sciences is therefore extended in important ways by Svendsen and her colleagues as they focus on what I would call "torques" (Bowker and Star 1999) or twists that disrupt the straightforward moral economy of sacrifice. If a research piglet was suffering and was not expected to live to the time point set by the experiment, the scientists would reluctantly but uniformly agree that it was better to euthanize the piglet (Svendsen et al. 2017, 210–11). If the piglet is killed, its sacrificial status is incomplete; it is less likely to achieve transcendence through datafication in that its death would no longer correspond with the end point delineated by the experiment. Janet articulated a similar kind of torque in the vignette above, when she stated that the animal facility would tell the lab that they would need to change their experiment and use the mice differently if the mice would not live to the time points set by the experiment. In both these cases, ending suffering is more important, but also puts into question whether or not that animal's life as a laboratory animal was worth living. Svendsen and her colleagues show that the naturalistic and analytic animal are therefore not discrete, and that the moral economy of sacrifice is not based upon the experiment alone. The moral economies of the experiment are braided together with the moral economies of animal sentience into a more uncertain moral economy of sacrifice.

In Svendsen's analysis of pigs as models in Denmark, emotions and affect are rarely seen. Svendsen et al. (2018, 68) state: "I had the confusing feeling that no existential issues had been at stake. Tiny piglets were born with pulses that made their bodies move up and down. Nonetheless, what happened appeared not as issues of life and death, but rather as a question of instrumentalizing biological life by connecting the piglets to other laboratory players and simultaneously detaching the animals from their species (the sow)." The emotions that arise in her fieldnotes when scientists are discussing euthanizing a piglet before the "kill day" are displeasure and a sense of hesitancy. The scientists do not express the kind of sadness that I experienced or that Janet described.

Indeed, Svendsen shows this by contrasting scientists' displeasure at killing a sick piglet early with Julie, an MSc student. New to working with laboratory animals, Julie becomes explicitly sad when the pigs she has cared for are being killed on the designated kill day of the experiment. We see in Svendsen's work that sadness gets lodged in the junior scientist, who then must learn how to manage her emotions so that sadness does not circulate. In arguing that sacrifice is not only a moral economy but also an affective economy, I want to suggest that the idea of sacrifice allows laboratory workers to learn how to not be overcome by sadness in the face of pervasive death. Sadness gets lodged and thus contained in individual bodies—in mine, Janet's, and Julie's—and this stops it from circulating through animal facilities.

We can think of laboratory animals that are killed without being turned into data as those who can be killed but not sacrificed (Agamben 1998). Sadness penetrates without recourse when sacrifice, as an affective economy, ceases to be a buttress. Svendsen and Koch (2013) and

Svendsen on her own (2022) turn to Derrida's analysis of sacrifice as the "absolute and mad contradiction between two responsibilities" (Svendsen and Koch 2013, S124). The absolute and mad contradictions that laboratory animals in science represent are between the simultaneity of the animal's subjectivity *and* its exploitation *as well as* the human exceptionalism that justifies the exploitation of animals *and* the connections between the human and animal that make such a substitution possible in the first place (Svendsen et al. 2017, 211). Sacrifice is not simply about killing in specific ways, but is also about negotiating the contradictory responsibilities that arise in doing science.

Sacrifice in science thus makes explicit the tensions between turning animals into data without creating too much suffering. Svendsen juxtaposes her ethnographic fieldnotes with Derrida's deconstruction of the Judeo-Christian-Islamic myth of Abraham with Isaac at Mount Moriah. Abraham acted responsibly toward God as the absolute other; scientists root their responsibility in universal knowledge. Abraham and the scientists do this by loving the one to be put to death (Svendsen 2022). Sacrifice is the bodily tension of these mad contradictions, wherein the fact of "animal use" creates intimacies between humans and animals. Pigs do not simply substitute for humans; humans must also substitute for pigs as caregivers. I argue that sacrifice makes this bodily tension—of killing and caring—possible, affectively.

The Sacrificial Logic in *Disgrace*

In her literary analysis of the novel *Disgrace* (Coetzee [1999]2010) and its specifically posthumanist ethics, Calina Ciobanu (2012, 668–69) traces David Lurie's statement that the dogs are "too menny" to Thomas Hardy's novel *Jude the Obscure*, in which the children are "too menny." Ciobanu argues that Coetzee is asking, through this reference, how it is possible to live within the specter of apartheid in postapartheid South Africa. Ciobanu notes that, in a humanist framework, this intertextual reference would be entirely inappropriate as it would suggest some kind of equivalence and commensurability between human children and dogs. But the point Coetzee is making, according to Ciobanu—and which I entirely agree with, as it is also the framework of this book—is that in questioning a humanist frame it is also possible to think ethics differently. How do we do ethics differently when animals are no longer viewed as mirrors to human hierarchies, but rather as actors who are also embroiled, albeit differently, within the situated frame of action? How are we bound up with others, human and nonhuman animals alike, in ways that may not be of our choosing, but to which we nonetheless must respond and ideally with some "ability" (Haraway 2008)?

The ending of *Disgrace* is a scene of "sacrifice." It raises questions about the irresolvable status of sacrifice in the face of disgrace (Ciobanu 2012). David Lurie, the novel's "disgraced" protagonist, finds some solace in playing the opera he is writing—an opera inspired by Byron—on a used banjo. A dog enjoys listening to Lurie play the banjo. The piece of music that was to be Lurie's final scholarly project has found an audience in a dog, who becomes Lurie's companion. When it comes time for this dog to be euthanized, the reader wonders if Lurie will keep this dog and give him a pass. But Lurie does not.

> He can save the young dog, if he wishes, for another week. But a time must come, it cannot be evaded, when he will have to bring him to Bev Shaw in her operating room (perhaps he will carry him in his arms, perhaps he will do that for him) and caress him and brush back the fur so that the needle can find the vein, and whisper to him and support him in the moment when, bewilderingly, his legs buckle; and then, when the soul is out, fold him and pack him away in his bag, and the next day wheel the bag into the flames and see that it is burnt, burnt up. He will do all that for him when his time comes. It will be little enough, less than little: nothing.
>
> He crosses the surgery. "Was that the last?" asks Bev Shaw.
> "One more."
> He opens the cage door. "Come," he says, bends, opens his arms. The dog wags its crippled rear, sniffs his face, licks his cheeks, his lips, his ears. He does nothing to stop it. "Come."
> Bearing him in his arms like a lamb, he re-enters the surgery. "I thought you would save him for another week," says Bev Shaw. "Are you giving him up?"
> "Yes, I am giving him up."
> (Coetzee [1999]2010: 219–20)

Ciobanu (2012, 685) argues that "there is no question that the novel's closing scene invokes the Pietà: Lurie as a Virgin Mary figure bears the Christ-like crippled dog 'in his arms like a lamb' and affirms with the last line ('Yes, I am giving him up') that we have stumbled on a scene of sacrifice." Ciobanu (2012, 220) asks if we are to interpret this as the "last" dog, just as Christ was to be the last sacrifice. Does Lurie's sacrifice redeem apartheid South Africa? But Ciobanu emphasizes that, given that Coetzee was so heavily influenced by Kafka, we must ask if this is an inversion of the sacrificial scene mirroring Kafka's (1919) "In the Penal Colony." Just as Kafka's Officer tries to sacrifice himself to preserve an old tradition and fails, so too does Lurie's sacrifice of the dog. Ciobanu argues that it is precisely the impossibility of resolving these two possible interpretations of the sacrificial scene that is the point. Irreconcilability, she argues, is at the heart of the posthumanistic ethic of the book. The posthumanist ethic aligns with Svendsen's (2022, 77) argument that: "What the interactions in the Newborn Pig Facility help us to see is that the framing that determines what falls within and what falls outside the norm of a human biographical, grievable life is constantly unsettled and transgressed by the very people who uphold it."

Therefore, the posthumanist framing of *Disgrace* can be used as a prism through which I ask the question: Was the sacrificial status of the geriatric mice I saw used in vaccine research worth it? I have had my two doses of a COVID-19 vaccine, built in part on the lives of these geriatric mice. Does the sacrifice of mice for my health, for our health, redeem the world? On the one hand, the answer must be no. COVID-19 vaccines reproduce global inequities, not only between species but also within species. I have had my two doses. I would have probably had four if I still lived in the United States. If I had lived in many locations under the label of "the global south," I would have probably had none. But if I hadn't had the COVID-19 vaccine, living as I do in the United Kingdom, what would that have accomplished? Quite possibly more deaths, more chronic illness, more domestic violence, and more mental distress. Each of these excesses, the "mores," would have reproduced existing social inequalities within the United Kingdom. Similar to Coetzee's final sacrificial scene, the question of laboratory animal sacrifice cannot be easily, definitively answered. Instead, I take from Coetzee the need to ask, whenever we find ourselves relying upon sacrifice, as an affective economy, to stop, pause, and feel some sadness.

The novel offers another possible sacrifice to diffract these questions. Lurie's daughter Lucy refuses to leave South Africa and the farm on which she was raped; she insists on keeping her child—despite the circumstances of rape. This sacrifice is something David Lurie cannot understand. Margaret Herrick (2016) has argued that Lurie's decision to sacrifice the dog shows his unwillingness to question his belief in abstract absolutes—such as sacrifice, the Romantic self as God, and, in turn, transcendence. David Lurie, therefore, consistently sacrifices the real world and the real bodies that are before him, including Melanie the student he rapes, and the dog whom he kills. Lucy, in contrast, refuses to escape from the world in this way; she refuses to run away from postapartheid South Africa. Herrick argues that Lucy is prepared to sacrifice an idealized version of the world. She accepts her embodied vulnerability in a manner that refuses violent retribution—a vulnerability that David Lurie claims makes her like a dog. Sacrifice in laboratory science, as a moral economy and an affective economy, engages in both these types of sacrifice—moving between the world as it is and a desire for abstract transcendence.

The Sacrificial Logic of Governing a Mass

We may be tempted to think that sacrifice, like killing, is a site where the lines between human and animal are retained as distinct. Animals can be sacrificed, but humans cannot and are not. However, as public health regularly calls upon people to make sacrifices, albeit not with their lives but certainly in their lives, this is only a partial truth. Quarantine during the COVID-19 pandemic was by and large construed and understood as a sacrifice that people made for the greater good. Sacrifice is therefore less of a divide between humans and animals, and instead another site of partial connection in the government of masses.

There are two important differences between the public health measures used within the laboratory and those used with people, however. In the animal facility, the individualized care that technicians provide to each mouse is tightly integrated with the population health measures the veterinarians implement. On the other hand, in global public health, care from a distance is prioritized and not well integrated with haptic medical care (V. Adams 2013). Second, when looked at through the prism of bioscience, there is no sense of transcendence when sacrifice is extolled in human health. Sacrifice in human public health is rhetorical, and it tends to be used to legitimize power relations and reproduce entrenched inequities.

Vincent, one of the veterinarians at the Institute, explicitly described himself as taking a public health approach to caring for the population of mice at the Institute.[7] Similarly, Alistair Anderson and Pru Hobson-West (2024) have shown that the focus on population health makes laboratory animals a unique site within veterinary medicine. The majority of veterinarians in Britain work in small-animal practices with pets. In contrast, the Named Veterinarian Surgeons working with laboratory animals, who were interviewed by Anderson and Hobson-West, described themselves as the "poor cousins" of veterinary medicine. They described themselves as "paper pushers" and "more hands off" when compared to "normal veterinarians." Vincent told me, "I cannot be the heroic veterinarian, running from sick animal to sick animal." I had assumed he was referencing James Herriot. This assumption is justified, as Anderson and Hobson-West found that the veterinarians they interviewed similarly mentioned James Herriot regularly. Herriot is a semifictional personification of a romanticized veterinarian. This image of the animal-loving veterinarian who works hands on with individual animals does not map onto the population-level, data-driven work that Vincent does on a day-to-day basis. He is more hands off, not (only) as a paper pusher but as a data analyzer. He cares for animals at a distance through not (only) bureaucratic mechanisms but also by making changes at the genetic or social-structural organizational level. He cares from a distance for a population, while the animal technicians are more likely to be focused on individual animals.

There is an asymmetry between medicine focused on the individual and public health focused on a population that creates moral tragedies for those working with laboratory animals, and this can also be seen in human health care. Katherine Mason (2016) has considered the dilemmas that arise when caring for a population by exploring the professionalization of public health as biomedical epidemiology in China.[8] She argues that Chinese epidemiologists had to become accountable to the transnational field of epidemiology. In the process, they risked sacrificing the local populations they were meant to serve in an attempt to integrate into the professional community. Mason shows how the question of who is served (e.g., the state, the people, global epidemiology) and who is governed (e.g., local populations) by public health is shaped by and shapes how "the commons"—or the community that shares resources—is imagined by practitioners. An asymmetry is created when the commons of professional public health is separated from the aggregate population being governed, or the community being served. This asymmetry raises serious ethical questions for public health measures in humanitarianism, according to Mason. Meanwhile, this asymmetry is the organizing logic undergirding the sacrifice of laboratory animals, in that the commons of public health is by definition focused on humans that are separate from the animals who can serve as human substitutes in research.

Quarantine is a useful site for exploring the trope of sacrifice in public health. The geriatric mice featured throughout this book are connected to human health as the test subjects for vaccines but also because they are quarantined in a cage within a biosecure facility as immunocompromised beings. Quarantine in public health also brings the sacrificial logic that operates through public health to the fore (Mason, 2016). Mason (2016, 19) demonstrates the problem of inequality that arises in this context:

> During an epidemic outbreak, for example, on behalf of the state a public health professional might very reasonably quarantine one group of people (for example, those who flew on a plane with a confirmed case of novel influenza) for the benefit of another (those who live in the city where the plane has landed). In one sense the quarantine represents public health privileging of the group over the individual, as the people on the flight are obligated to forego their individual freedoms temporarily so that the city might remain disease free. . . . But even at the level of the group we can see that there is an imbalance of cost and benefit: One group (those on the plane) is sacrificing for another (those in the receiving city). The group being quarantined will not necessarily benefit from the quarantine, even at a group level; in some cases, most or even all members of the group might be harmed. . . . Still, in this case, protect-

ing the unexposed group is deemed more important. The group being targeted and the group being protected are necessarily not one and the same. What may make this seem fair and reasonable is that the distinction between the group that is served and the group that has to sacrifice is in theory circumstantial and temporary. . . . Those individuals making up the group being sacrificed in the name of a larger common good, the theory goes, may well benefit the next time, when others might need to sacrifice their freedoms to protect them. . . . But this only works if the chances of being in the sacrificial group versus the benefiting group are equal for all—a proposition that rarely plays out in real life.

Public health often requires one group to sacrifice themselves, not in the form of certain death as seen with laboratory animals, but in the form of physical restrictions and potential exposure to a novel virus that contains the risk of death. This sacrificial logic is considered reasonable when the sacrificial group feels that they are part of the "commons"— and could just as well benefit from such sacrifices in another circumstance. Inequities in the chance of being in the sacrificial group instead of the benefiting group undermine this logic, however. Mason (2016, 183) notes that, for younger members of the professional epidemiological middle class in China, "self-sacrifice became something that others were primarily responsible for doing." This attitude toward sacrifice is not unique to China but rather to the transnational middle class, where young Chinese professionals have been taught to aspire—and the reference point is largely the United States.

Moreover, the idea that "self-sacrifice" is something other people are responsible for is certainly not unique to China's professional epidemiological middle class. Writing from Britain, this is more than abundantly clear, as investigation after investigation finds that former prime minister Boris Johnson and those in his party held and attended parties while the rest of the nation was in lockdown. Sacrifice is something that the British ruling political class called upon other people to do, which, when looked at in this way, isn't sacrifice at all. There is no transcendence nor universal ideal. Sacrifice here is instead a rhetorical obfuscation for the exertion of power and domination. Looking back at the laboratory animal from this prism is important; sacrificing laboratory animals is also a site of species domination.

More-than-Human Humanitarianism and Sacrifice

The use of animals in science is given a sense of purpose and transcendence through sacrifice. A good life and a good death for a laboratory animal is one where suffering is kept to a minimum and where the animal body is transcended as timeless data that can create a common good—albeit for another species—through a planned life and death. Sacrifice is also an affective economy, one that helps to keep sadness at bay, relying on it being contained within individual bodies and not circulating through the lab. This containment means that scientific research with animals, day in and day out, is made possible by a sense of purpose. Without the affective economy of sacrifice, the extent of animal death that human health relies upon can become overwhelming. After all, I willed myself to stop crying, to bury my emotions, so that I would not create distress for the mice who were being killed and the technicians who were having to do the killing.

To conclude, I want to consider how the moral economies of sacrifice are currently mutating. While Lesley Sharp (2019) found that the majority of scientists in her research project spoke of "animal sacrifice," veterinarians and technicians did not. Mette Svendsen (2017, 63) found that scientists she interviewed in Denmark disliked the term "sacrifice." I rarely heard scientists, veterinarians, or technicians use that term and saw it used far more in Tarquin Holmes's historical research. This historical research provides a potential prehistory to Lesley Sharp's analysis of the word "sacrifice" in science. Sharp (2019) states: "'Sacrifice' might best be thought of as a 'Key Word' (Williams, 1983) whose complex history enables us to track shifting ideas about the medical, social, economic, and moral worth of using animals in scientific research." She finds the first use of the term in a scientific journal to appear in 1903. Building on this idea of sacrifice as a keyword, it is worth noting that the term did appear frequently in the 1875 Royal Commission. John Simon, who supported vivisection, argued that scientists should be "anxious not all to underrate the real fact that the life of the animal is sacrificed for physiology" (Holmes and Friese 2020, 49). Meanwhile, Brad Bolman (2018) found extensive use of the term "sacrifice" in his analysis of the mid-20th-century reports of UC Davis's Radiobiology Laboratory that used beagles as model organisms. Does the once prominent, but now seemingly declining, significance of sacrifice as an actor's category possibly allow us to see a way out of sacrificial logics?

There is an emergent idea of creating "cultures of care" in laboratory science, which I would posit is replacing the idea of sacrifice. Cultures of care are a relatively new organizational morality, which has developed in response to high-profile failures of care (Nuyts and Friese 2023; Gorman and Davies 2020). Cultures of care explicitly cut across human health care and laboratory animal facilities and seek to create points of connection between laboratory animals and patient representatives (Gorman and Davies 2020). There are diverse ideas about what constitutes a culture of care, but going above and beyond legal requirements in providing care for caregivers and care receivers is a key goal (Greenhough and Roe 2018). The abstractions associated with sacrifice and transcendence are declining in this context as new forms of governance and managerialism come into being. Killing animals persists in this milieu but without the religiosity, inwardness, and sense of universal transcendence that is associated with sacrifice. A culture of care could replace the role of sacrifice as an affective economy. But if a sense of purpose is not rooted in the timelessness of data, in the universality of beneficent knowledge, what kinds of affective economies hold such a culture of care together? Within animal facilities, the affective economy of a culture of care tends to be rooted in a sense of belonging within the workplace. Is "belonging to a workplace" a strong enough affective economy, one that can hold up, day-to-day, to the question of whether or not it is worth it for animals to die in order to promote human health?

Humanitarianism through the Prism

What does this analysis contribute to humanitarianism and human rights? I have focused on the sacrificial logics of science in partial connection but also differentiation from the sacrificial logics of public health. Meanwhile, humanitarianism and human rights scholarship have explored the sacrificial logic of the law (Van der Walt 2005). Kafka's "In the Penal Colony" has not only inspired Coetzee's posthumanist ethic but also much of the theoretical work on the sacrificial logics of the law by Derrida, Agamben, and Walter Benjamin (Van der Walt 2005; Liska 2022). This scholarship has sought not only to expose sacrificial logics but also to theorize a way out of the sacrificial logics of the law. In the context of critical legal studies, Johan van der Walt (2005) suggests, through his analysis of sacrifice in the work of Derrida, that courage may be such a path where courage is a "vulnerable and precarious strength to live the uncertain life of mortals." This courage sounds to me a lot like the path that Lucy in *Disgrace* takes. In order to get out of sacrificial logics, the weight of history in the present must be accepted such that we are all subjugated *and* refuse to participate in the mechanisms of violent retribution. The lesson that can be learned from Lucy is to accept the world before us by forfeiting—literally cutting out, as she cut Lurie from her life—the abstractions of transcendence. Through this prism, humanitarianism can ask what kind of affective economy can hold up to the weight of history, and love the world without violent retaliation.

5

Compassion

Compassion for another may *feel* like a natural response to their suffering, but compassion can be better understood as learned. One of the things that I realized while conducting this research was how beloved rats are by many scientists and technicians. On my first visit to an animal facility, I saw Elspeth's appreciation for rats as highly intelligent and sociable animals and how this informed her understanding that tethered rats are suffering. While most of my participant observation was conducted in Elspeth's laboratory and then at the Institute—I also visited several other facilities. On one of these short-term visits to another university animal facility, I spent much of my day seeing how the animal technicians created games to play with the laboratory rats. The joy they expressed in their inventiveness and the sheer fun of playing with the rats was palpable. As anyone who has a pet rat will attest, it is possible for a human to engage in a dyadic relationship with a rat that includes cuddles and play. When interacting with mice, this is simply not the case; they are deeply skittish in all interactions with humans unless something is seriously wrong. For example, while an animal technician told me that she *does* get to know the mice she works with as individuals, this knowledge is developed at more of a distance with mice when compared to rats. Within laboratory-based practices, forming a compassionate relationship with a suffering rat can be easier than with a mouse. I must admit that I struggled to engage with the mice I observed and instead had to learn to become compassionate toward them.

This chapter tells the story of how cages of mice, an animal technician named Janet, and a sociologist became "alongside" (Latimer 2013a; Bennett 2020) one another on a specific day through the partially connecting experience of being replaceable. I use my specific emotive response within this relational situation as the basis for the analysis, and show how I learned compassion in the process.[1] I can only speak to my experience of that moment and cannot be sure what Janet or the mice felt. But, as Matthew Newcomb (2007) argues in his attempt to recuperate compassion from Arendt's critiques of this emotion: "When compassion is known to be limited and impure in matching the feelings of another, it can become a conscious act of imagination (an attribute valued greatly by Arendt) that feels alongside another, but never fully identifies with that other."[2] As such, this is not a story of equivalence; a mouse, an animal technician, and a sociologist did not collapse into one another.

The feeling of being replaceable was the "affective economy" (Ahmed 2014, 2004) that made compassion possible. This chapter, therefore, unpacks and questions this feeling of replaceability. I do this by amplifying my feeling of replaceability by discussing the novel *Washington Black* (2018) by Esi Edugyan. I then situate this feeling of replaceability vis-à-vis the idea of "replacement" that is one of the 3Rs—replace, reduce, and refine animal experimentation—and an ethical mandate for scientists who use laboratory animals. For laboratory mice, the only means of escaping their lives within bioscience is to die or to be unborn by being replaced.[3] Henry Buller (in Schrader and Johnson 2017, 11–12) puts it this way concerning farm animals: "They exist to be killed, their lives are stage-managed to arrive at a pre-defined killing. . . . There are two means of escape: to die . . . or to be unborn. . . . And yet, to be killable is also to be alive, to have lived." Looking at replaceability through this lens emphasizes this state of being.

In the humanitarian literature, compassion is normally understood as the starting point—the trigger for action, intervention, and social justice. In presenting compassion as an outcome rather than a starting point, this chapter shows how compassion can instead be woven together as a tool of solidarity.[4] Rather than asking what is owed to laboratory animals, which is often the starting point for an animal justice framework (e.g., Nussbaum 2023), I conclude that a more-than-human humanitarianism seeks out ways to be in empathic and solidaristic relationships with animals. In developing this argument, I draw on Barbara Prainsack's work both alone (Prainsack 2018) and with Alana Buyx (Prainsack and Buyx 2017) on solidarity. Solidarity is a distinctly relational ethical and political concept that Prainsack and Buyx argue counters the idea of the bounded, independent, and rationally self-realizing individual. Interdependence is treated as a social fact. Wherein animal justice is aspirational in demanding people accept the costs of what they owe to others, solidarity begins by looking at those relations where people are already willing to accept costs to assist others with whom they recognize some trace of similarity. I argue that both animal technicians and scientists experience this recognition of similarity without sameness, just as I experienced this recognition. They are willing to accept certain costs to assist animals who live in a world that may not be as they want it to be but is the world that they are confronted with.

Vignette

Janet and I spent our day together in a room within the animal facility where two other animal technicians—Martine and Sarah—also worked with the mice. The room was arranged with several rows of cages jutting from one wall and two fume hoods against the opposite wall. Sarah was coming in and out of the room throughout the day. Martine worked with the mice on one side of the room and at the neighboring fume hood. She was directly beside Janet and me throughout the day. Retrospectively, I realized that Janet and I were having semipublic conversations throughout the day. On reflection, I think that we each performed our roles in that room: Janet performed as the animal technician, and Carrie performed as the visiting sociologist.

However, toward the end of the day, we found ourselves in a quieter setting. I followed Janet as she cleaned up and moved between various rooms of the experimental unit of the animal facility. She brought the used equipment to the "dirty room," which was silent and empty, with no windows looking into it. Metal shelves surrounded the edge of the room, and it was extremely antiseptic in feeling. There was only one door leading into the room, and there was another double metal door behind which the dirty equipment was left. The room was white and metallic; there was very little else in it. Inside this separate and quiet room, Janet leaned back against a metal shelf. She was silent for a moment. I followed Janet, and leaned against the facing metal shelf and waited.

Janet then started to talk. She told me that her family knew she was an animal technician, but none of her friends knew what she did for work. I was not immediately surprised by this revelation. I knew she had been an animal technician for over 30 years and would have worked in this profession during the late 1990s when some animal rights activists were taking violent measures. I also knew from other sociological research that animal technicians frequently do not publicly announce their profession for fear of judgment (Michael and Birke 1994; Birke, Arluke, and Michael 2007; Arluke 1991). I continued to listen.

Janet went on to explain that animal rights activists had targeted her in the late 1990s, and her family was threatened as well. When her family was threatened, Janet initially decided to quit her job. Again, I was not surprised, and I can imagine feeling quite the same. That one would want to protect one's family from violence is assumed by many; this is precisely why this kind of tactic is used. What did surprise me was Janet's reasoning for not quitting.

Janet continued to tell me that she knew she would be replaced. She worried that her replacement would not care about the animals as much as she did. And so, she decided that she owed it to her animals to stay in the job, to continue to look after them. That is why she is still an animal technician today.[5]

I had a physical reaction to this part of Janet's story that is difficult to put into words. I was surprised by Janet's decision, and it felt like the story moved through my body. I also felt a very strong emotional reaction, and I still do to this day. The emotions are probably best described as a mixture of incredible respect and a degree of sadness. The feelings I experienced were profound; I was moved by Janet's story and commitment to her work, as well as her care for and about laboratory animals.[6] After a moment of quiet, Janet got up and went to the door. She had made a confession of sorts and was ready to move on. I had to pull myself together and quickly follow her; I had to move on just as Janet did and was.[7]

* * *

I think what makes me feel not only respect but also sadness when I think about Janet's story is how she makes explicit a shared vulnerability with laboratory animals. I could imagine this vulnerability because I have felt it myself. Janet seemed to understand herself as *being perceived as replaceable*—at least by the organization where she worked in the late 1990s (which is not the organization I was shadowing her at) but probably more generally by the institution of science. Someone else would be hired in her place. And they might not be as good at the job, at least from the perspective of the animals—whose perspective could never be fully known but that Janet was centrally concerned with. This trope of being replaceable has an affinity with the laboratory animal. Laboratory animals are valuable as a population—or at least laboratory mice are. Individual mice are largely understood as interchangeable and thus entirely replaceable (see Sharp 2019).

The laboratory animal that is seen to represent the scientific potential of the species is distinguishable from a pet based on their replaceability. Individuality and uniqueness are at particular risk in workplaces, and this is what is thought to distinguish the public from the private space embodied by the home.[8] In his analysis of people's relationships with their dogs, Clinton Sanders cites a phenomenological psychologist who says this about his dog:

> History informs the experience of a particular animal whether or not it can tell that history. Events in the life of an animal shape and even constitute him or her. . . . Sabaka is an individual in that he is not constituted through and I do not live toward him as a species-specific behavioural repertoire or developmental sequence. More positively, he is an individual in that he is both subject to and subject of "true historical particulars." . . . *I can not replace him, nor ethically, can I "sacrifice" him for he is a unique individual being.* (emphasis added, Shapiro 1986 in Sanders 2003, 410)

Lesley Sharp (2019) points out that "sac-ing" is commonly used to describe the killing of laboratory animals in the United States. She notes the double entendre where "sacrifice" is shortened to perform the workplace nomenclature of "being fired" or "sacked." Both Janet and the mice are replaceable in the workplace, as most of us are as workers.

Janet seemed to express that she cares about the laboratory animals by being open to sharing the vulnerability of being replaceable with the laboratory animals she cares for. What is expressed—or at least what I heard—is a relatedness between animal technicians and laboratory animals where nonpower is experienced as shared across species. There was nothing I could do but hear this and thus also experience this nonpower. But in the process, I felt like I could also start to imagine the life of a laboratory mouse. I, too, am also replaceable in the workplace. In order to unpack this moment where I learned to become compassionate toward laboratory mice through my compassionate response to Janet, I turn to the theme of replaceability.

Replaceability as an Ethical Mandate

Replaceability has thus far been discussed in negative terms. Still, we should remember that replacing laboratory animals with insentient models (e.g., computer simulations, tissue cultures, yeast, or—possibly—less sentient animals like flies and worms) is an ethical mandate for scientists to engage in. It is the first R in the 3Rs—replace, reduce, and refine animals from science—which is, albeit contested (McLeod and Hartley 2018), the transnational gold standard for doing scientific research with laboratory animals. The goal here is to replace distressing techniques using animals with techniques that do not use animals and thus replace animals completely from some areas of scientific experimentation. While the absolute replacement of animals from science is held up as the ultimate ideal and many resources are invested in this goal, refinement and reduction are practically the primary compliances with the 3Rs. However, replacement is the goal, and it is one that scientists and animal activists have historically agreed upon (Myelnikov 2024; Kirk 2018).

Russell and Burch published *The Principles of Humane Experimental Technique* ([1959]1967) over 60 years ago, and here introduced the concept of the 3Rs.[9] The research and resulting book were commissioned by UFAW (the Universities Federation for Animal Welfare) following their successful publication of the *UFAW Handbook on the Care and Management of Laboratory Animals,* primarily concerned with the husbandry work of animal technicians and focused on the time the animal spends outside of the experiment. Russell and Burch's study was instead primarily concerned with the work of scientists with animals in the experiment itself, albeit with the recognition that experiments take place over days, months, and even sometimes years—making it impossible to separate out husbandry from the experiment. The 3Rs were developed to "diminish inhumanity in experimentation," specifically in creating a scientific subject who is humane to the animals used in science. A key distinction Russell and Birch make is that a great many procedures used in science need to be either "totally free of direct inhumanity" or "may be taken as of only slight direct inhumanity, involving the sort of distress which human blood donors cheerfully accept" (Russell and Burch [1959]1967, ch. 4b). A painless death of an individual animal is an ethical solution in this context. But the idealized death of a strain of laboratory animals, through the replacement of one scientific procedure with another, is the more idealized solution.

The word "compassion" is not used in *The Principles*, but (in)"humanity" and (in)"humane" are repeatedly used across the book. Compassion has much in common with these words; both refer to kindness, care, and sympathy toward others, especially those suffering. In the English language, we have tended to use "compassion" when referring to a human emotion toward other humans, whereas "humane" is used when referring to other-than-human animals. Human exceptionalism is built into the language, shaping the sentimental structure of our emotions. In the process, compassion is naturalized as a distinctly human phenomenon.

Replaceability

I have thought a lot about the conversation I shared with Janet. I believe that there was a movement away from interpolating one another as our professional roles (e.g., animal technician and sociologist) to engaging in a process of narrating and listening. It was an interactional process that Lisa Stevenson (2014) has called song. If interpolating puts someone squarely within their social identity, Stevenson (2014, 162–65) asks and responds:

> Can we imagine listening or speaking to someone without fixing her identity in advance? . . . This is what I want to call song—seeking someone, calling someone, singing to someone, and perhaps, yes, even interpolating someone (if we can dissociate the term from its roots in hating)—as company, as a presence. . . . Song, as I intend the term, draws attention to forms of address that seek the company of an other rather than those that attempt to identify, situate, or render an other intelligible.

At this moment, I think Janet was narrating herself to me in a song that sought company in the workplace. In the process, I came out of my role as the visiting sociologist and became Carrie—a person who could imagine Janet's commitment to her work. We were in each other's presence as people rather than in our respective roles. A shared sense of replaceability in the workplace allowed me to imagine what it was like to be Janet.

The horror of replaceability is central to the narrative in *Washington Black* (2018) by Esi Edugyan. The novel addresses—among many other things—the horror of being replaced by one you love and the guilt one feels in having replaced one who has loved you. *Washington Black* is an adventure novel that is structured by the horrors of slavery, British colonialism, and imperialism. It is also a love story. While there is a romantic love story within the novel, most of the novel explores love in asymmetrical hierarchies: George Washington Black (aka Wash) as a child to his mother, Big Kit, and Wash as a slave to Titch Wilde. These two loves structure the adventure of Wash's life. We see three individuals here trying, at times succeeding and at times failing, to break out of how they are interpolated vis-à-vis one another.

Erasmus Wilde is a cruel slave owner who has taken over Faith Plantation in Barbados—a profitable site for the British Wilde family. Erasmus's overt and despotic violence toward the people he enslaves, and toward his family as well, sets the frame within which all the actions of the novel must respond. There is no way to escape Erasmus's violence, even after Erasmus dies. In showing how Erasmus's grotesque cruelty overdetermines all other action, constraining what any single person can do, Edugyan spends most of her novel dissecting the smaller violences that occur as people try to survive. Surviving is an act done both alone and together when faced with the cruel and despotic violences of colonialism, slavery, and racism. In the process, Kit, Titch, and Wash himself exert smaller violences and injustices upon those they love. Replaceability is one of those violences.

The first moment of replaceability comes when Wash replaces his mother, Big Kit, with Titch. Wash is transformed from a fieldworker who lives side by side with Big Kit—his mother—to a house servant qua scientific assistant who lives side by side with Erasmus's brother Titch. We learn that Titch is an abolitionist and a scientist; Wash becomes his research assistant. Titch teaches Wash to read, and they both discover that Wash has an exceptional talent for drawing, which is particularly useful as a research assistant to a natural historian. Wash has found the safest harbor he could probably have found in plantation slavery, but he also must do what Titch "says and not ask why." Wash has no contact with Big Kit during this time, and although he misses Kit tremendously, he lives in fear of returning to the fields.

Months later, Wash finds himself at Erasmus's dinner table again, but this time he is attending to Titch and not serving. There are two field slaves serving dinner—an older woman and a young boy—just as he and Kit had served dinner months ago. The older woman had been seriously maimed, presumably as a form of cruel punishment; Wash had also suffered serious burns to his face from a scientific experiment. The older woman flashes Wash a smile, and we know she recognizes him. Wash, however, only recognizes the older woman as Kit at the very end of the dinner, as she and the young boy are leaving.

> How could I not have known her? Had I not all these months prayed for her deliverance each night, imagined for her a life beyond the blood-blackened fields of Faith? . . . She was much changed, it was true, maimed terribly, grown thinner, the hair at her temples silver as flies' wings. Aged, now, as though decades had separated us. But I was the more changed; that was the uglier truth. I gripped anxiously at my hands, staring at Kit's tall figure. How solicitous she was with the boy. I saw she kept a careful eye on his posture, his manners. I knew instinctively what this meant, the great angry love she held that boy inside, like a fist. (Edugyan 2018, 99–100)

It is the last time Wash sees Kit, although he remembers her, senses her, and feels her presence for the remainder of the novel. He knows he replaced Kit and that she tried to replace him as well.

Following a suicide that Wash is made to witness, Titch and Wash are forced to flee Faith Plantation in their experimental "cloud cutter" that is essentially a hot air balloon. As the pair travel by boat to the Arctic, Erasmus puts a bounty of $1,000 out on Wash. In the Arctic, they find Titch's father at his scientific outpost. Wash's loyalty to Titch never wavers, and while he is not enslaved anymore, Wash is also not free. Titch gives Wash his freedom in the Arctic by abandoning him to the freezing snow of the scientific outpost. Wash initially assumes that Titch committed suicide but later learns that he survived. Titch's survival makes the line between freedom and abandonment, love and self-interest all the more ambiguous.

For the remainder of the novel, we see how Wash builds a life for himself after slavery but in the violence of racism. Sometimes, that racism is brutal and physically violent. For example, Wash meets the bounty hunter, who—despite knowing he can no longer claim the money—nonetheless tries to kill Wash. Sometimes, that racism is brutal and institutionally violent. For example, Wash must give credit for his invention of the aquarium to his girlfriend Tanna's father—who is a white natural historian.

Wash is haunted by both Kit and Titch, and needs to know what happened to both in order to feel able to build his life. His struggles revolve around how he replaced Kit with Titch and, ultimately, that Titch replaced him. Upon learning that Titch may be still alive, Wash recounts his life in slavery to his girlfriend Tanna; it is a story of the monstrosity of replacing Kit to escape the horrors of slavery. Later, Wash states: "How was it that I had lately given more thought to the possibility of his [Titch] being alive than to Big Kit's death? It was shameful. But my sense of betrayal shook me deeply—the idea that Titch had cut, rather casually, my tie to him, which was all I'd had in the world" (Edugyan 2018, 321). Upon having it confirmed that Kit had died on Faith Plantation as an "apprentice" if not a "slave," Wash remembers her love and his turn from that love.

The cut goes both ways—Wash turned from Kit to Titch, and Titch turned from Wash. But we cannot quite dislike Titch for abandoning Wash, just as we cannot dislike Wash for abandoning Kit. Even Titch's biggest critic, Tanna, cannot help but be kind to him. When she meets Titch in Morocco, with another young boy serving as his scientific assistant, she still feels sympathy for Titch's own pain. Titch tells Wash, "I did not want you to think I had merely replaced you" (Edugyan 2018, 402). But in a sense, Wash learns that Titch had replaced him because these two men were running from two very different horrors: Wash from the horrors of slavery and Titch from the horrors of his own brutality. Titch had not directed this brutality toward Wash but toward his cousin, whose suicide Wash was forced to witness.

The novel foregrounds how replaceability coincides with one's relationship to labor. Labor interpolates people into their roles, and this becomes the basis for all kinds of hurt that occurs in the resulting relationships between people. However, although we are told of the acute and monstrous suffering Wash experienced as a slave, the novel itself is about the psychic suffering that results from this but does not determine Wash's person. In this way, Edugyan addresses the horrors of slavery without the risk of slipping into a gratuitous depiction of suffering. I take this to be an important starting point for considering how compassion can address and meet suffering in practice without interpolating a whole person—or, in the case of my own research, an animal—within a role. Rather, in *Washington Black*, compassion is uncomfortably built into the relationships, the beings and becomings alongside (Latimer 2013a; Bennett 2020), of people in survival.

More-than-Human Humanitarianism, Compassion, and Solidarity

Focusing on practices of compassion can extend what Arlie Hochschild (2016) has referred to as "empathy maps," which are individual and collective ways of distributing empathy across social relations, where some are deemed worthy of high empathy, others of low empathy, and some fall within a no-empathy zone.[10] *Washington Black* (2018) creates such an empathy map. Edugyan leads both her characters and, in the process, her readers to feel compassionately for people they may not have been prepared to empathize with.

The image of the animal rights activists would be one who tends to have high empathy for laboratory animals and no empathy for scientists and technicians. Proscience advocates are countered as having high empathy for scientists, low empathy for laboratory animals, and no empathy for animal rights activists. Is it possible to counter this polarization by having a high empathy level for laboratory animals, animal technicians, veterinarians, scientists, and animal rights activists? Is it possible to recreate here what Natalia Ruiz-Junco (2017, 425) calls an "empathy path": tracing togetherness that can create compassion for a widening set of actors to reduce the numbers of those deemed outside of compassionate feeling?

Would it be better to never live than to live in science? Can a life in science ever be "a life worth living"? Replacing a mouse strain is more or less the same as making it go extinct. This tension runs through *The Principles* itself. On the one hand, Russell and Burch state: "Our domesticated species are swept up in our own triumphal career. Among these are laboratory animals, and William Lane-Petter has pointed out to us that, if experimentation can become completely humane, we may be doing these species a considerable service in keeping them alive" (Russell and Burch [1959]1967, ch. 3a). Yet on the other, they also rejoice at how tissue culture presented, in the 1950s, a way to completely replace some animals from experiments, referencing Sanders who

> concludes in general that tissue culture methods provide (in the nontechnical sense) more information, more precise information, and new kinds of information; he also predicts with confidence that this type of replacement will continue unabated. . . . "The animal virologist has great cause to rejoice at his liberation from the hazards and uncertainties of animal experiment." (Russell and Burch [1959]1967, ch. 5d)

I do not wish to argue against the ideal of entirely replacing animals in science, but I think it is important to point out that this requires not only the death of animals but also many having not lived, to have been "unborn." To find oneself replaceable is to have at least lived with another and, possibly, compassionately.

Prainsack and Buyx's (2017) concept of solidarity is particularly helpful in this context. They argue that solidarity is an aspirational practice of relating while also being agnostic regarding normative concerns, such as whether or not laboratory animals *should* exist. Prainsack and Buyx argue that solidarity can be an alternative model for arbitrating ethical dilemmas across various areas of social life. I here use it as a model for health and medicine wherein well-being is a social achievement rather than an individual responsibility or trait. Solidarity counters the idea of the bounded, independent, and rationally self-realizing individual with an alternative, relational understanding of human beings. If relations and interactions rather than independent individuals are the starting point, solidarity stages ethics within current practices rather than focusing on an idealized world to come. This allows us to ask questions such as how to respond to the animals now on the planet who can only live as laboratory animals—remembering that most laboratory mice are immunocompromised and, therefore, cannot live outside of science.

Prainsack (2018, 25) summarizes that solidarity starts with the fact that "human relationality is a precondition for subjectivity, not the other way around.... We are who we are because we relate to others." More-than-human humanitarianism shares this understanding of relations as the ontological basis of the world. Importantly, Prainsack and Buyx (2017) define solidaristic practices as those in which people are willing to accept costs in order to assist others with whom they recognize some form of similarity. Janet certainly accepted costs to assist laboratory animals, and her actions can thus be seen as compassionate and solidaristic. There is an elective affinity between compassion, more-than-human humanitarianism, and solidarity.

Melanie Rock and Christopher Degeling (2015, 432) have extended Prainsack and Buyx's conceptualization of solidarity to include other-than-human species.[11] They begin with the social fact that many people—and in this I would include Janet—already feel solidarity with other-than-human species and environments. Rock and Degeling note, however, that there is a particularly weak link between public health ethics and animal welfare ethics, which they highlight by drawing attention to zoonotic diseases. In this context, Rock and Degeling suggest that a more-than-human solidarity approach is important because it does not bracket, stamp out, or underplay how animals are entangled not only in human states of health and illness (through zoonotic diseases) but also in therapeutic responses (through laboratory animals). And this importantly raises the issue of conflicting solidarities. They give the example of the horse owner who exposes themself to an equine disease as an attempted act of solidarity with their horses, but who then shifts the potential health burden to veterinarians instead. A focus on more-than-human solidarities thus pries open the "selectivity" (Wahlberg and Gammeltoft 2018) of solidarity and forces the question of contesting needs to the fore.

In this context, it is worth highlighting Prainsack's argument that one of the outcomes of a solidaristic approach to ethical questions in medicine is the need for universal health care. Health care, as currently understood by either Western biomedicine or traditional Asian medicine, is heavily reliant upon animal bodies. Therefore, we cannot expect solidarities to easily align, as more health care in the present seems to imply more reliance upon laboratory animal lives and deaths. Recognizing the full range of trade-offs and burdens that are being made allows incongruous interests to come to the fore rather than remain hidden. It is crucial for either justice or solidarity approaches to humans or animals to retain empathy paths in highlighting these trade-offs and burdens, as compassion and solidarity can just as quickly be transformed into resentment and sacrifice. The current historical moment makes this abundantly clear. More-than-human solidarity is thus one aspiration that can arise through the practices of more-than-human humanitarianism.

I have often wished that I had said something that would have indicated to Janet how much her approach to her work moved me. While I was moved, emotionally and internally, I didn't *do* anything. I was silent. I "hesitated" (Lopez-Gomez 2019) probably because there was nothing I *could do*. Compassion was felt, but it did not result in any particular action and even led to inaction. Compassion, however, was an outcome that allowed me to expand my empathy paths to include mice.

Humanitarianism through the Prism

One of the characteristics of humanitarianism is an ethos of empathic response to suffering, attending to and holding on to individual experience within a mass of anguish, such that one is driven to act to end another's torment (Newcomb 2007; Bennett 2020). Compassion for a distant other is central to the "affective economy" (Ahmed 2004) of humanitarianism.[12]

This "sentimental structure" (Sharp 2019) in humanitarianism is one that animal activists have frequently called upon to create compassion for animals—whether the theoretical basis of that activism is utilitarian such as Peter Singer (1975), poststructuralist such as Henry Buller (2013), or abolitionist such as PETA.[13] Compassion is linked to justice or a concern with redressing what is owed to another. The linkages between empathy, seeking to help another, and seeking to redress suffering through justice can be seen in Martha Nussbaum's book *Justice for Animals* (2023). Nussbaum embodies the suffering that all humans cause to animals (which includes me and most readers of this book, who may not be meat eaters but who certainly live in human-dominated landscapes) in individual animals. The goal is for that individual story to prompt a feeling of empathy that can translate into action in the form of legislative change. Our empathic response to the particularity of each of these animals is the basis for her call that humans, collectively, give animals back what is owed to them. Compassion, justice, and rights are thus tightly woven together.

What we can ask of humanitarianism, through the lens of more-than-human humanitarianism, is how compassion and justice are linked in a way that is taken for granted. Compassion can be configured more broadly to include solidarity and the expansion of empathy maps and paths that are more inclusive. The compassion that I felt and describe in this chapter was an *outcome* of togetherness rather than a starting point for togetherness. It was not motivated by what anyone would call suffering. The mice I saw that day were not suffering in any acute sense, nor was Janet. Rather, the psychic toll of being replaceable in a working environment that becomes a life—the entire life of the two-to-three-year-old mice (which, it should be emphasized, is *very* old for mice), 35 working years in Janet's life, and at that point, 15 working years in my own life—is the site of partial connection. *Feeling alongside* without fully identifying becomes possible when *being alongside* (Latimer 2013a) . It is a difference-making imaginative process that resists the pull toward interpolation.

In this sense, my hope is to engage in another definition of compassion that has dropped out of use. Marjorie Garber (2004) notes that the word "compassion" meant both suffering together as fellow feeling and emotion felt on behalf of another who suffered from the 14th to the 17th century. The former definition, however, fell out of favor.[14] With this, Garber argues (2004) that the creation of an "abject other" who is offered assistance is both the proudest boast and undoing of compassion; the risk is that this creates the power relations and hierarchies of pity. I argue that we need to recover the older definition of the word "compassion" in order to address this risk, to recognize moments of compassion as an outcome of being in the presence of another—as opposed to understanding compassion as the starting point for the action of helping another. The practices of compassion, not its "promised" outcomes, should be the foci (see also Kienzler 2019). More-than-human approaches to humanitarianism highlight this older definition of compassion and show how compassion can be done and felt when fellow feeling is prioritized, not as a bounded feeling within an individual who has the power to choose to give to another.

Compassion is as an important social accomplishment.[15] It is a privilege to experience compassion. And this is where humanitarianism can take stock. What happens when compassion is not viewed as a necessary starting point for action, but rather as a privileged emotional response to the action of being together? This emotional response may (or may not) arise from being alongside others—human and nonhuman animals as well as other kinds of lives and things—one would not ordinarily get to be in the presence of. Creating conditions to be alongside those you wouldn't normally find as company is therefore both a necessary ambition, and also a serious sociological challenge.

6

Consent

A mouse does not consent to participating in research.[1] The presumption of a rational, individuated human actor who can consent to research or therapeutic intervention upon their body and self means that extending the theme of consent to laboratory animals is, seemingly, questionable at best. But if we understand consent as a *relational* process, this extension of consent can, indeed, be helpful—not least for addressing the many dilemmas that consent gives rise to within biomedicine itself and in the relationships that are enacted between differently situated humans.[2] Indeed, the question of consent in animals can be seen as a long-standing concern among ethologists, and was a crucial question that Gregory Bateson ([1972]2000) posed in "A Theory of Play and Fantasy." Within the context of play, Bateson asked not how a monkey says "yes," but rather how a monkey says "Don't" in a manner that is heard and acted upon by another. Inspired by Bateson, this chapter recounts how I witnessed an animal technician hear a mouse say "Don't."[3] I contend that the question of how to hear a "Don't" is crucial to discussions about consent, as coercion can occur when a nonlinguistic "Don't" is ignored in the context of a linguistic yes.

Consent is also crucial to social research, including my own. Joanna Latimer and I conducted ethnographic research alongside one another at the Institute, wherein I spent time in the animal facility and various laboratories using the mice and Latimer spent time in laboratories using "insentient" organisms such as yeast, flies, or worms. When gaining entrée into the Institute as a field site for conducting ethnographic research, one of the questions that arose was how we would gain consent from the individual scientists and technicians we were shadowing. Scientists in Britain are often skeptical of allowing social scientists into their workplaces, and concern over the violent tactics used by animal rights activists from the 1970s through the 1990s is often the rationale for this. However, individual scientists and scientific organizations also understand the importance of being open and transparent about their work. On the one hand, because taxpayers are funding much of the basic scientific research that is conducted, scientists are increasingly required to engage with the public as a condition for receiving this funding. Many scientists, however, would also like to present their work and how they care for animals to outside audiences. In a context where the organization felt compelled to work with two social scientists and some individual scientists were also compelled, how would we ensure that those scientists who didn't want to participate in our research were also left out? How would we recognize and leave out scientists who didn't want to be in the research? We worked with the Institute and our respective institutions' research ethics committees to address these questions. Going through these consultative and bureaucratic processes was crucial.[4] But the question of research ethics always also exceeded these formal ethical procedures (Latimer and Puig de la Bellacasa 2013). This chapter examines one of the ethical excesses, focusing on how we hear another say "Don't" without language.

Informed consent is foundational to biomedicine as practiced with humans, both in medical/therapeutic and scientific/research practices. Informed consent ensures that a person is not coerced into receiving treatment or being researched upon without express permission. As important as informed consent is, the practice has been critiqued extensively for a range of reasons, including the fact that it takes an individualized approach to what is often a social process (Fox and Swazey 1984; Corrigan 2003; West-McGruer 2020; Prainsack 2018).

The question of consent is also present but more tendentious in the context of humanitarianism. In his history of Euro–American humanitarianism, Michael Barnett (2011) asserts that paternalism is a potentially necessary but nonetheless problematic characteristic of the relationship between givers and receivers of aid.[5] Humanitarian aid often presumes that the receiver of that aid wants help, and that consent is therefore implied (Stevenson 2014). In this context, I suggest that one of the key benefits of more-than-human humanitarianism is in making consent an ever-present question. How more-than-human humanitarianism seeks to hear nonverbal articulations of "Don't"—a skill that must be cultivated over time and through the development of intimate knowledge—is an important lesson for humanitarianism and social science researchers alike.

Vignette

The animal technician who I am following today is cleaning cages. Sarah has been doing this type of work for over 40 years, most of her adult life. She explained that she grew up close to an animal facility and started doing this kind of work when she was young. And so here she is—working in a different animal facility but doing the same type of work decades later. As a result, Sarah has worked with almost every kind of laboratory animal across her career.

Sarah takes a metal cart and puts four cages on the bottom and four on the top shelves. She pushes the cart to the hood. Sarah disinfects her gloved hands, takes a cage, and places it within a rectangular metal device for opening the cage. She takes off the cover of the cage, observes the mice, and quietly talks to the mice. She then apologizes: "I'm sorry I talk to my mice. I know they can't understand me, but I can't stop. I think it calms them also, as it is stressful for them to be disrupted like this." Mariam Motamedi Fraser (2015) has made words a distinctly sociological object of inquiry, asking what is done with words in interaction and what can be learned by probing nonlinguistic word relations. Sarah's talk with the mice is a place where words flourish in a "non-discursive assemblage," mirroring those explored by Fraser (2019, 131). It is not the content of what Sarah is saying to the mice that matters. Indeed, this is why she apologizes to me for talking to the mice, indicating a possible sense of embarrassment.[6] Rather, making the sound of words with one's mouth while physically handling the mice was a bodily, sensorial, and affective experience that I suspect Sarah found relaxing herself. I know that I found it relaxing to watch and listen to her work.

Sarah walks away to get a clean cage with preprepared sawdust and bedding. She disinfects her hands. I watch the mice run around the cage, somewhat surprised that they do not try to escape with the lid being open. They don't, which Sarah clearly knows. One by one, Sarah moves the mice to the clean cage, inspecting their bellies along the way, bringing some old bedding and older wood chips into the clean cage and the red tunnel. She also includes some fresh food and sunflower seeds as a treat. While Sarah noted that cage cleaning is stressful for the mice, it also seemed to me that she used the cage cleaning as an opportunity for play and a time for the mice to explore as they move from one cage to the other. Stress is not necessarily bad and can be highly productive and necessary; deep play for children is an example of a productive type of stress. For mice living in a cage for so long, some stress is probably necessary, and cage cleaning seems like a good opportunity for this play.

Sarah carefully recovers the clean cage, places it on the metal cart, puts the dirty cage on another rack, and begins the process again. She turns to me at one point and says, "Are you sure you won't get bored?" I tell her I won't, that there is something relaxing about her work. Her talk with the mice is part of what makes watching this work relaxing: the repetition of movement that has a kind of grace combined with words spoken for their sound rather than their content. There was something meditative about watching Sarah work, which was like watching a dance.

Like any job rooted in movement, animal technicians' work is physical labor, but the grace of it can make this easy to forget. It is physically hard to be on one's feet all day, to bend up and down repeatedly to get cages, and to fit one's body within the hood all day long. Sarah's shoulders are a bit hunched forward, and I wonder if this is age- or work-related. Her body looks perfectly fitted within the hood as if her work has shaped her body to fit its parameters. But the animal technician's work is not only physical labor. It certainly cannot be automated in the way that this animal facility has automated the cleaning of the physical cages with robotics. The rhythm and routine are ruptured at key moments.

At one point, for example, Sarah suddenly turned the light off over the hood before moving a cage to it. She quietly said that one of the mice seemed about "to fit."[7] She was going to do this cage quickly. I was quiet and allowed Sarah to focus on cleaning the cage. After it was done, I asked how she knew the mouse would fit. Sarah responded that after all these years, she knew the visual cues, the bodily freezing, and when an animal was about to fit. Sarah knows through patterns, habits, and similarities, such that when things are different she also knows there is a problem (see chapter 2). So, Sarah got the new cage ready and moved the mice as quickly as possible, as this was not a time for play for these mice. Sarah is habituated to the signs of productive versus unproductive stress among mice. She knows when a mouse is intimating "don't."

How Do Animals Say "Don't"?

Sarah's use of play and her ability to hear mice say "Don't" as part of that play is a social accomplishment. Gregory Bateson ([1972]2000) famously found it curious that monkeys act as if they are fighting while playing and bite each other as if fighting without producing pain or animosity. Bateson ponders: "What I encountered at the zoo was a phenomenon well known to everybody: I saw two young monkeys *playing*, i.e., engaged in an interactive sequence of which the unit actions or signals were similar to but not the same as those of combat" (Bateson [1972]2000, 179). It is through Bateson that I came to understood Sarah's work in both allowing mice to play and hearing mice say "Don't" to that play as significant and in need of appreciation.

To further understand Sarah's work as a site of consent, I note that Eduardo Kohn (2013) has developed Bateson's analysis of the metalinguistics of play. Kohn has done this to develop his more-than-human model of thinking, a relational model of thought not located in the brain but in interpretive and embodied interactions that always include other-than-humans. Kohn extends Bateson's analysis of monkey play to the dog play he witnessed as part of his ethnographic research. For Kohn, the key question is how monkeys or dogs can say "Don't" in the context of play. Kohn (2013, 147) states:

> Whereas negation is relatively simply to communicate in a symbolic register, it is quite difficult to do so in the indexical communicative modalities typical of nonhuman communication.... Saying "don't" symbolically is simple.... But how do you say "don't" indexically? The only way to do so is to re-create the "indexical" sign but this time without its indexical effect. The only way to indexically convey the pragmatic negative canine imperative, "Don't bite".... Is to reproduce the act of biting but in a way that is detached from its usual indexical associations.

Drawing on Bateson and Kohn, I argue that the fact that Sarah can hear the mouse say "Don't" indexically must be considered an accomplishment in the context of this scholarship on language and communication.[8] And being able to hear "Don't" instead of a "yes" or a "no" seems crucial for consent to not become coercion.

A mouse fitting during a cage cleaning is not the same kind of conundrum as a dog biting another dog during a play fight. When a dog or a monkey bites another in an intraspecies encounter, "Don't" is expressed by the indexical sign of acting out the thing that is not wanted; this is what distinguishes it as play. On the other hand, a mouse fitting is interpreted as a "Don't" in the context of an interspecies encounter. A mouse fitting is a more stable signifier of "Don't" to Sarah than a dog play-biting another dog. "Don't" is expressed by the mouse to Sarah, but not by making the indexical sign of that which is not wanted as in the case of the dog biting another dog during play. Nonetheless, being able to say and hear "Don't" during play is at the heart of both Bateson's and Kohn's conundrum and Sarah's work.

As Bateson said, the very proposition of play requires that the animals in question are "capable of some degree of meta-communication, i.e., of exchanging signals which would carry the message 'this is play'" (Bateson 1987 in Motamedi Fraser 2019, 133). To understand Sarah's work—in giving the mice a space to play during the cage cleaning while also allowing them to say "no" to that play—I believe we need to understand words as productive beyond their meaning. On the one hand, this is shown in Sarah's use of words as calming to say aloud, regardless of their content. But equally this is shown when the word "Don't" is expressed without language. Miriam Motamedi Fraser's (2015) analysis of words beyond language, and her extension of this analysis to dog words (2019), helps me to understand how animal technicians and mice communicate with one another. I therefore describe Motamedi Fraser's theory of more-than-human and more-than-linguistic word relations in order to better understand how Sarah listens to the mice, and communicates with them without language.

* * *

In developing dog words conceptually, Motamedi Fraser's (2019) primary interlocutor is the dog trainer Vicki Hearne and her notion of dog gestures. In developing dog gestures, Hearne interprets Bandit's gestures as communicating beyond a stimulus–response model, wherein sit = biscuit.[9] Rather, Hearne shows how a sit can mean different things within a particular milieu or interaction, such that a sit could mean "I want to work," "I need to go out," or "I want to eat." Motamedi Fraser extends this to argue that these actions are not only dog gestures but also dog words, in that a dog sitting is a signifier to which no stable signified is attached.

Motamedi Fraser emphasizes the physicality of making a word, whether it be with one's mouth or hand (which we associate with humans) or with one's body (which here we associate with dogs):

> A dog makes a dog word by shaping her or his body into a particular position, or by moving her or his body in a particular way (in the way of a retriever for example), just as a spoken word is made by the sound that is shaped by the face, mouth, larynx, etc. and a written word is made by the movement of the hand and the marker. A dog makes a word with her or his body, and this is all that the dog makes, no more and no less than a word. (Motamedi Fraser 2019, 137)

A dog sitting makes an image, and that image is interpreted and acted upon by others rather than having a stable signified attached.

Motamedi Fraser makes this argument by showing how her dog uses his body near the refrigerator to intimate that it is time to eat. She articulates this understanding of their communication as follows:

> I am not suggesting that a dog word has no meaning, or that a dog does not mean what she or he says. On the contrary, *the conceptual dimension of the sign is suspended only until the dog word arrives at its meaning by way of the relations that constitute it*. These relations are not with other words—to be blunt: Monk uses words, but he does not make sentences—but with close-to-hand and familiar resources that enable a dog to intimate. "To intimate," Berlant writes, "is to communicate with the sparest of signs and gestures, and at its root intimacy has the quality of eloquence and brevity" (Berlant, 1998, p. 281).
> (Motamedi Fraser 2019, 137)

Building upon Motamedi Fraser's argument, we could similarly say that the mouse fitting is a mouse word. The mouse's body speaks a word when it starts to fit. This word is interpreted by Sarah as the mouse saying "Don't" to the playtime of cage cleaning. The mouse fitting intimates "Don't" through the relations that constitute it, including Sarah's previous encounters with mice. Sarah hears this "don't," and her habit shifts; the play stops as this would not be productive play for the mouse. The cage is cleaned as quickly as possible, so the disruption of the cage cleaning is minimized. By cleaning quickly, Sarah says, "Don't fit" to the mouse.

What is important here is that to hear a dog word or a mouse word, wherein bodily words are signifiers to which there is no stable signified is attached, an intimate relationship is required. Motamedi Fraser states:

> A dog's gesture *can* be understood as a word, if a word is understood as part of an "associative complex" or encounter. Inventing associative complexes or participating in word encounters is a way of making and using words (non-referentially), and also a way of thinking (concretely). In an associative complex, meaning is generated when any number of ostensibly "disconnected" entities are brought together on the basis of "perceived similarities, commonalities, or relationships." While the bonds that define such relationships will necessarily be many and varied (Monk and I, for instance, especially with our different *umwelts*, identify similarities very differently), they will always be situational, and specific. It is for this reason, I wager, they will also intimate. (Motamedi Fraser 2019, 139)

Dog words and mouse words are not located in the individual animal but in the "intimate entanglements" (Latimer and Lopez Gomez 2019a) through which those bodily signs make (or don't make) worlds. There is a real danger in the fact that these words can be ignored because they take so much embodied knowledge to hear. "The stakes in a word encounter are thus very high, for failing to acknowledge or respond to potential meaning is likely to lead to the profound injustice of stifling thinking" (Motamedi Fraser 2019, 143). Word relations make dog worlds, and I would add mouse worlds.

Word relations are relationships of power that make the world but are also made of the world. Word relations are constrained by the conditions of possibility for listening well, not only through the interpretation that is based on previous representations of the world but also by the conditions of possibility for responding. As Greenhough and Roe (2019, 374) make clear: "If we learn anything from the ATs [animal technicians] . . . it is that the infrastructure imposes an inability to escape animal exploitation, and AT attunement to animals is part and parcel of that."[10] With a relational understanding of mind as distributed across bodies rather than located in brains through Kohn, combined with attention to power relations through Motamedi Fraser, it needs to be emphasized that Sarah does still clean the cage, even if she hears the mouse as saying "don't." Sarah must abide by the law and the terms of her employment. Sarah cannot respond to the mouse's word by not cleaning the cage even if she wishes to. She can only react by cleaning the cage entirely differently, shifting from what looked like a "playful" encounter to a calm but quick encounter. The start of a fit in a mouse, bodily freezing as a sign, could significantly stand for "Don't play" to Sarah; "Don't clean," however, was not an option.

Amplifying Consent

Animals do not figure significantly in Edugyan's novel *Washington Black* (2018), except for one scene. This scene is, nonetheless, important in marking a key transition of protagonist Wash's attachments. Wash moves from a period of detachment and isolation, of "bitter self-sufficiency" after being abandoned by Titch, to turning toward Tanna as a love interest. Turning toward Tanna necessitates attachment to Tanna's white father, Goff. Tanna asks Wash to dive, in her place, for a living octopus that will become a dead specimen for her naturalist father to bring back to Britain for display. Tanna cannot dive because of a broken wrist. While Wash does not "like to feel ill-used" by Goff, he nonetheless finds himself doing the dive as it allows him to be close to Tanna. He consents, but that consent is sticky.

Diving was a dangerous practice, and its danger was overdetermined by racism and colonialism. Goff tells Wash, "Well, the main thing is to try not to die. I shall give you some advice on how best to bring that about" (Edugyan 2018, 270). Free diving is a long-standing practice in many parts of the coastal world but became an exploitative and deadly practice in the context of the British Empire. In *Coral Empire* (2019), Ann Elias shows how aquatic environments were racialized and coastal people subordinated as part of British colonialism in both the Bahamas and Australia through fishing as part of industry, collecting as part of natural science, and filming underwater as part of entertainment. The fact that Goff uses his mixed-race daughter in the first instance to dive, and asks Wash to dive in Tanna's place, must be read in the context of this racist and colonial history.

Wash succeeds in collecting a female octopus during his dive, and he does this by engaging with the octopus and getting her "to like him."

> The octopus arranged itself in a smatter of algae, its body hanging blackly before me. When I came forward to touch it, it sent out a surge of dark ink. We paused, watching each other, the grey rag of ink hanging between us. Then it shot off through the water, stopping short to radiate like a cloth set afire, its arms unfurling and vibrating. There was something playful in the pause, as if it expected me to ink it back. I held my hands out towards it, gently; the creature hovered in the dark waters, almost totally still. Then, shyly, it began to pulse towards me, stopping just inches away, its small gelatinous eyes taking me in. Then it swam directly into my hands. (Edugyan 2018, 271)

Wash's meeting with the octopus is not an instance of animalization, which would create an equivalence between Wash as a racialized person and the octopus as an objectified animal. Rather, the fact of Wash and the octopus being together—underwater, at that moment alongside one another—is embroiled in the history of racialization and colonization that made "animalization" possible. Wash's compassion for the octopus is what makes collecting her possible. Edugyan here creates an "image" (Stevenson 2014) of Hannah Arendt's provocation that compassion creates inequality and Mary Douglas's proclamation that compassion is repression. Both Arendt and Douglas were famously critical of compassion as an emotion that cements inequalities through a pretense of helping another. Here, we see how compassion can be instrumentalized in coercion, which looks like consent.

Wash first comes up with the invention of the aquarium during this dive, and in response to the octopus. "When I thought of Goff killing her to crate up as a specimen for his exhibition, a twist of nausea went through me. How wrong it all felt. Could she not, I thought, be brought to England alive to be seen as the breathing miracle she was?" (Edugyan 2018, 271). Here, Wash imagines the aquarium, an institution that developed in parallel with colonialism; the objectification of marine life parallels the objectification of colonized people (Elias 2019). Nigel Rothfels

(2008) has shown how immersive zoo exhibits were derived from the practice of exhibiting people in ways that contributed to their racialization. Rothfels argues that the illusion of personal freedom generated support for these exhibits among professional anthropologists and the general public, even though these people were treated as property, as servants and slaves. Wash does not picture the aquarium as making the octopus potentially free, but rather as at least allowing the octopus to stay alive. But we are left to wonder: Would staying alive be better for this octopus than her imminent death? Wash is a survivor, and for him, the answer to the vexed question—is this a life worth living—is always clear: staying alive is better. However, the reader is left to ask what Wash himself is complicit in as he imagines the aquarium while in this moment of being alongside this octopus.[11]

Vignette

I am having lunch with a group of scientists who all work in the immunology lab, where I have observed research on aging and vaccine uptake. They are all either PhD students or postdocs and are relatively junior within the hierarchy of science. We are sitting outside on a picnic bench in the shade created by the animal facility. I feel somewhat self-conscious at this lunch. Our picnic bench is below a window I have looked out of before while having lunch with an animal technician. I know that this is the kitchen window for the section of the animal facility where I have spent much time conducting fieldwork. I felt strange having my lunch outside and below this window with the scientists who used the mice these technicians cared for and eventually killed. It was a sunny day, and I felt guilty for having lunch outside while the technicians were having their lunch indoors.

This feeling of being in the shadow of the animal facility takes on new meaning as the scientists begin to discuss the possible uses of the green landscape that surrounds us. They wonder why this park-like setting isn't being used for recreation and start considering the different sports that could be played here. One suggests croquette, and someone else jokes that they could use the mice as the balls. They are clearly referencing *Alice in Wonderland* ([1865]1993) here, replacing the hedgehog balls used in the game of croquette with laboratory mice. Everyone laughs. I am smiling, too, thinking that Derrida also turned to *Alice in Wonderland* in *The Animal That Therefore I Am* (2008). One of the more senior scientists in the group sees me smile at this moment, and maybe she even sees me thinking. I see her see me smile. I also see her shift quickly in affect from relaxation to seriousness. She says to the group, "As said in front of the visiting sociologist." Everyone stops laughing; the topic is changed.

I believe that this scientist was worried about me telling this story because it could convey the wrong impression: that these scientists don't care about their animals and are even possibly cruel to animals. I agree with her that this would be the wrong interpretation—if this is what she was thinking. I am certain that these scientists do care about animals. My research to date has shown scientists by and large care for and about animals above and beyond regulatory requirements, not least in order to be a good scientist. I think the joke here represents the kind of humor that Lesley Sharp (2019, 131–32) has shown occurs routinely in clinical medicine, scientific laboratories, and research animal facilities where young professionals must learn how to become accustomed to suffering and death.[12] These scientists will be with suffering and death, day in and day out, for the whole of their careers. Young professionals must learn to live in the daily presence of painful emotions that these events give rise to. A macabre humor is a long-standing way for young scientists and clinicians to express their ambivalences and discomforts within this challenging milieu. The joke does not indicate an uncaring attitude but rather a very real personal and professional struggle that occurs when one is hoping to do good, and that requires doing some real harm.

I interpreted the scientist as telling me: "Don't tell this story." It is a "Don't" that I have listened to, and I have deleted this section of the book many times only to put it back in. My dilemma is that I did not interpret their jokes as indicating an uncaring attitude. I am listening to this scientist, responding to the "Don't" I heard in her affect by telling the story differently. But is this the right decision? None of this would fall under the rubric of formal consent, as I received informed consent to do this research ethically. But this is the kind of ethical practice that informed consent as a process necessitates, particularly in the context of ethnographic work conducted in an organization where people consented to allow me to shadow them for the day and opted out by not talking with me while I was shadowing someone else in their laboratory (see Plankey-Videla 2012). I cannot know if this is what she was actually thinking, just as I am certain she did not know that I was thinking about Derrida and his analysis of hedgehogs when I heard her say, without words, "Don't tell this story."

And maybe I am wrong. Maybe she, too, knows that Lewis Carroll was an antivivisectionist who saw secularization, science, vivisection, and selfishness as one and the same (Collingwood 1898, 167–70). She worries that I will make too big a thing out of this joking reference to *Alice in Wonderland* as part of the playfulness that is necessary for interpretive analysis.

Ethics and Poetics

I would like to suggest that these scientists turning to Lewis Carroll, just as Derrida turned to Lewis Carroll, may help extend the more-than-human theory of knowing and of hearing dissent without words. To understand what it takes to hear "Don't", we need to understand it as poetic: the impossible double bind of hearing a "Don't" as part of the visceral experience of presence. Presence defies language, but nonetheless somehow needs to be put into language. I will start with how the hedgehog helps us to understand poesis through Derrida before turning to what this means for hearing "don't.".

In *The Animal That Therefore I Am*, Derrida has this to say about *Alice in Wonderland* and the hedgehogs:

> Although I don't have time to do so, I would of course have liked to inscribe my whole talk within a reading of Lewis Carroll. In fact you can't be certain that I am not doing that, for better or for worse, silently, unconsciously, or without your knowing. You can't be certain that I didn't already do it one day when, ten years ago, I let speak or let pass a little hedgehog, a suckling hedgehog perhaps, before the question 'What is poetry?' For thinking concerning the animal, if there is such a thing, derives from poetry. (Derrida 2008, 7)

Derrida provides a genealogy for his thinking on animals for philosophy by tracing hedgehogs in *Alice in Wonderland*—a story that both resists and reproduces the Cartesian duality of mind/body and human/animal—to his use of hedgehogs to think poetics. For Derrida, hedgehogs provide a way through the difficulty of grasping, linguistically, the metalinguistic. Hedgehogs did for Derrida what monkeys playing did for Bateson.

One of the important elements of the hedgehog is that it challenges binary thinking about wholes and parts. In this sense, this genealogy also resonates with the concept of "partial connections" (Strathern 2004) and "being alongside" (Latimer 2013a). Timothy Clark (1993, 46) notes that Derrida is not the first to reference "the lowly hedgehog" in reflecting on poetics.

> By choosing to relate the poetic in terms of the fable of the hedgehog, Derrida is engaged with one of the best known of the Athenaeum Fragments (1798), no.206. Friedrich Schlegel writes there of the fragment form in relation to the idea of a transcendental poetry, a poetry that will have no empirical referent but would exist as the essential or the absolute poetic creativity (poesis) itself. A fragment, in relation to such an ideal, would be paradoxically 'like a miniature work of art', 'entirely isolated from the surrounding world and . . . complete in itself like a hedgehog.' For Schlegel the hedgehog-fragment, in its isolation, its would-be perfect closure upon itself, figures the Ideal of the Romantic text as unconditioned, i.e. absolved of relation to anything but itself—*ab-solute*. (Clark 1993, 46)

There is a whole, in the sense of complete and isolated, definition of poesis that Derrida critiques by reclaiming the hedgehog. To be both whole and isolated is a romantic ideal. As a side note, this is also the romantic ideal that David Lurie in J.M. Coetzee's novel *Disgrace* is motivated by and that the novel itself critiques.

However, Derrida argues that the hedgehog only rolls up into a whole in relation to another who is perceived as dangerous. According to Derrida, this exposure to distress, to finitude, is poetic experience. Clark analyses Derrida as follows:

> Correspondingly to hearken to the poetic is to attend to something, both in and beyond language, which is not 'just as it is' but whose identity is

unstable, wild, impatient of definition: a mode of alterity not of sameness. It thus cannot be known: knowledge as such must be renounced if one is to perceive this creature that, unexpectedly, crosses the thoroughfares of communication, from one side to the other. The poetic moves or disturbs obliquely; its space is one in which the language of identification and exchange is traversed by what cannot be simply said or thought. (Clark 1993, 50)

Derrida thus reclaims the hedgehog by challenging that it is not an idealized whole that is pure knowledge, for it cannot be whole without relation, and this relation exceeds the linguistic.

What does the double bind Derrida focuses on through the hedgehog mean when attending to hearing another say "Don't" without language? Clark's description of the poetic articulates how I felt in the presence of Sarah as she quickly dimmed the light of the hood in which a cage of mice was positioned, or the senior scientist as I thought I heard her intimate 'don't tell this story'. Sarah doesn't know what the mice are thinking, nor do I. I don't know what Sarah or the postdoc were thinking. The postdoctoral scientist didn't know I was thinking about Derrida. I was moved, though, and I have tried to put that movement into language. It is a needed but equally impossible task. Playfulness is my interpretative strategy, allowing me to link Derrida's hedgehog with Bateson's monkey and Kohn's and Fraser's dogs.

Timothy Clark analyses Derrida's meditation on poesis through the hedgehog as an ode, an address or an appeal that speaks to a future that may never be, a future that language anticipates. "The ode may thus work performatively, giving itself as an act of dedication, worship or supplication" (Clark 1993, 53). To value the work of hearing "Don't," we need more odes to lowly creatures like hedgehogs and mice who are always part of relationally producing knowledge. Maybe that ode is the work of the visiting sociologist and distinguishes her from the ethicist.

More-than-Human Humanitarianism and Consent

Consent relies upon spoken and written language and can be formalized with a "yes," often by signing one's name. However, hearing another say something that we may take as a "Don't" relies upon more than language. Hearing a "Don't" requires intimate knowledge (Raffles 2002). How might we recognize the work that goes into hearing "Don't"? How might this ability to hear and respond to a "Don't" be recognized and valued without trying to codify it in the process? To address these questions, I started by exploring the theme of interspecies communication and relational knowledge. Then, I moved on to the vexed question of how to hear another say "Don't" without words. These moves trouble the focus on individual freedom and self-ownership, which are often taken for granted in the processes and procedures of informed consent. The hedgehog is a countering image of consent and coercion, as always in something of a dialectical relationship to one another in the everyday interactions of life.

It should again be emphasized that laboratory animals do not consent to being in biomedical science. However, it is worth noting that both informed consent and the regulation of laboratory animal welfare stem from the belief that rules and regulations help keep tyranny at bay.[13] For example, the impetus for British regulation of animals in science in the 19th century was exactly this. It was egregious practices occurring as part of vivisection that, in part, prompted the Cruelty to Animals Act of 1876. Since then, a range of practices—including the 3Rs (replace, reduce, and refine animals from research), the scientization of animal welfare, and "cultures of care"—have all sought to address everyday ethical concerns. I agree with Tone Druglitrø and Kristin Asdal (2024) that there are different versions of care, and this includes not only the type of "skilled care" (Druglitrø 2018) described in this chapter but also "procedural care" (Druglitrø and Asdal 2024; G. Davies et al. 2018), which includes paperwork, bureaucracy, and the creation of standards. However, I would like to suggest in this chapter that laboratory animals show us both the need for and the limits to a rule-based approach to ethics—highlighting what is needed in addition to bureaucratized procedures like informed consent. I make this argument by asking the following question: What intimate knowledges are required so that a person in a relative position of power can detect, hear, and respond to nonverbal signs that intimate "Don't"?

If formal ethics has given us informed consent—where the individual understands and says yes to their proposed future—virtue ethics provides us with the challenge of hearing and responding to a "Don't" that is always also not a "no." Formal and virtue ethics need one another in this context; neither can be fully relied upon. The problem is not formal ethics per se, but rather how its formality can render invisible and thus unnecessary all the virtues of engaging in the hard work of listening well. A more-than-human humanitarianism puts these needs at the forefront of both formal and informal ethics. Time and embeddedness are needed to cultivate the skill of being in a position to receive consent and dissent.

Humanitarianism through the Prism

I would like to conclude this chapter by suggesting that the work and time that goes into interspecies communication—and the attendant critiques of languagism presented in animal studies by anthropologists like Eduardo Kohn (2013), sociologists like Mariam Motamedi Fraser (2019, 2015), and feminist ethologists like Vinciane Despret (2004, 2005, 2008, 2013, 2016)—is important for humanitarianism to consider. Can recipients of aid say "Don't" in a meaningful way that is listened to and responded to? Do aid workers have the intimate knowledge gained through an extended time spent day in and day out with the receivers of aid that is required to hear a "Don't"? A focus on language as a form of communication and the codification of consent could make it even more difficult for humans with very different experiences in and of the world to say and hear "Don't" to and from one another. Indeed, in their study of consent in the context of combat sports, Channon and Matthews (2022, 910) similarly find that "the formalization of consent within the rules and norms of a sport undermines athletes' perceived need to purposefully engage with consent as a process of clear and explicit interpersonal communication."

I will give an example from social studies of humanitarianism here to demonstrate how consent comes to be assumed and the dangers that arise. Lisa Stevenson (2014) describes what forms of anonymous care can mean to a person who is subjected to a humanitarianism that presumes a biopolitics rooted in life. She explores this topic ethnographically, through an empirical focus on the tuberculosis epidemic in the mid-20th century and the suicide epidemic of the early 21st century among native people colonized by the Canadian state. For Stevenson, the suicide hotline is an exemplar of anonymous care. Originally developed by the Good Samaritans in London, anonymity was viewed as crucial to this intervention that sought to save lives. Common humanity was the basis for a staged interaction between a volunteer and a caller who did not know one another. The telephone connects two who share a humanity, allowing for care through language. Stevenson is careful to note that anonymity, of course, has benefits, particularly in small communities. However, she shows how legacies of colonialism persist, partly because an assumption is made that biopolitics rooted in preserving life is good and in the best interest of people. Consent is, therefore, generally assumed. Stevenson questions what it feels like to be made to live regardless and irrespective of who one is. In other words, what does it feel like to be an index of the proper workings of another's humanitarianism (e.g., to be one who has been saved from suicide)?

Despret (2016) asks, as an ethologist, "What would animals say if we asked the right questions?" When taken seriously, the counterintuitive nature of the question and the hard work it requires could serve as an important prompt for humanitarians. Are humanitarians asking the right questions of the receivers of aid? What might be elicited if different questions were asked? Based on Monika Krause's (2014) sociological examination of humanitarian projects as work that is greatly defined by the field, I would like to suggest that the answer is often no. Krause shows this is not because of pigheadedness but because managers need to produce "good projects" as a commodity, and the beneficiaries are part of that commodity form. Beneficiaries are competing for aid, and in order to receive resources, compliance with that commodity form is mandatory (Krause 2014, 37, 52). Humanitarian project managers and beneficiaries risk losing resources by asking different questions or responding to unexpected resistances—even if those questions might yield a better humanitarianism. The parallel in this chapter is Sarah, who must change her practices of cage cleaning rather than stop cleaning in response to hearing a mouse say "Don't" because of the institutional field within which she works.

To close, I want to return to Miriam Motamedi Fraser's point that: "The stakes in a word encounter are thus very high, for failing to acknowledge or respond to potential meaning is likely to lead to the profound injustice of stifling thinking" (Motamedi Fraser 2019, 143). Motamedi Fraser was here talking about dog words, signs to which a stable signifier is not attached. Humans, however, are also always trying to make new words to make new worlds; listening well becomes a way to attend to and participate in these worlding practices, in the flow of life with one another rather than according to a prescribed index. Based on illness narratives that were elicited during her ethnographic fieldwork with Kosovar Albanian women from 2007 to 2009, Hanna Kienzler (2022) builds upon scholarship on the elusiveness of communicating pain to ask what relations are needed for bodies to speak in and through pain. She articulates this as "SymptomSpeak [which] allows people who perceive each other as similar enough to co-articulate truth claims" (Kienzler 2022). How did Kienzler, as a dissimilar ethnographer, engage with SymptomSpeak? It was a language or words that Kienzler had to learn to hear as part of the worlding practices of an ethnographer involved in the flow of life, relationally.

Conclusion

I am spending the day shadowing Adam, a postdoctoral researcher conducting a series of experiments that ask why older people respond less well to vaccines when compared to younger people. Adam explains to me that it is known that older people do not uptake immunizations as well as younger people do, but the mechanism behind this is not known. His experiments aim to understand those mechanisms better. He explains that immunizations were given to mice at 12 weeks of age and 90-plus weeks of age. Seven older and seven younger mice were then culled at different time points after the immunization was given: 7, 10, 18, and 21 days.

Today I am watching the last experiment, using the seven young and seven old mice culled at day 21 after their immunization. When I arrive at the laboratory, Adam's work is well underway. He had taken lymph nodes from the 14 mice and is now mashing them through a very fine mesh sieve, held over a petri dish, using a saline solution. Adam adds more saline to each petri dish and puts the petri dish against a little motor that runs it in circles. He puts the petri dishes in a centrifuge and adds a culture before putting the cells in an incubator for about 20 minutes.

Adam refers to these moments as "waiting for culture," and during this time we chat about topics that include but also exceed science. He tells me how he likes to organize his weeks by ordering experimental work and data analysis. We speak about being non-British academics in the UK and how Brexit brings new uncertainties to our careers. We talk with his lab partners about everything from politics to exercise routines. Because of the division of labor in science, everyone I spend time with in this laboratory who is studying aging is significantly younger than me. They are working hard to carve out a future for themselves in scientific research while being concerned about and cognizant of how an uncertain global geopolitics will impact them. These were often informal conversations, where humor played a role in a marked casualness that was entirely different from how these young scientists talked about their experiments. These conversations were a way to pleasantly pass the time as we waited for cells to be cultured.

* * *

It was Adam's term—"waiting for culture"—that inspired me to explore the interactional moments of participant observation by juxtaposing them with fiction. I have sought to describe some of the affective dimensions of more-than-human humanitarianism practiced with laboratory animals today, placed in historical context. The merits of fiction were to make the seemingly mundane visible and, hopefully, valuable. While Adam and I would wait for a culture that amplifies DNA so that it can be seen, I have used fiction to amplify what may have otherwise been fleeting, ephemeral moments; this brought the sociological significance of these moments to the fore. Where Adam's culture is a liquid made up of nutrients, my culture is novels made up of stories.

"Waiting for culture" is partially provocative as a heuristic because there is a historical narrative that we still live with today, in which science is understood as a "culture of no culture" (Traweek 1988). Politicians' claims during the COVID-19 pandemic that they were "following the science" drew on this narrative and reinstated this idea that science is outside of the contaminating forces of society and can be easily followed as a neutral path outside of politics. Early work in science and technology studies (STS) was premised upon troubling this narrative. Michael Lynch's (1989) canonical conceptualization of laboratory animal sacrifice was a move to understand science as a social activity—one that can be understood or studied just as sociologists study any other field of social life. The idea that we were waiting for culture seemed to evoke this idea of science as a neutral space outside of culture and social life.

Yet, simultaneously, what makes the idea of "waiting for culture" humorous is the way the phrase "waiting for" evokes Samuel Beckett's *Waiting for Godot*. Like Vladimir and Estragon, Adam and I spent our time waiting for culture by discussing various topics that allowed us to pass the time but did not seem significant in the context of the universal scientific truth we were waiting for. The immunological research that Adam was conducting often eluded me, and I spent that day desperately trying to understand his research. During this culturing process, I could lose the thread of the science and no longer understood how the mouse lymph node I saw in the morning was connected to the graph I saw in the evening. Our conversations about exercise routines seemed to me entirely beside the point.

After rereading these fieldnotes many times, I realized these conversations were the point. *Waiting for Godot* is not simply a comedy but a tragicomedy that is explicitly about ethics. Elliott Turley (2020) defines Beckett's ethics in *Waiting for Godot*: "To live ethically is not to emerge from a dilemma but to enter a fix. *Godot* draws on the collective wisdom of an array of philosophical and literary thinkers, but it insists on staging that wisdom across the muddled throes and failures that constitute a lived ethics." Waiting for culture helps articulate the simultaneous and seemingly contradictory ways in which Adam and I were in the thick of culture, in the thick of lived ethics, there in the laboratory. The universal, transcendent knowledge we were waiting for, which never came—to me at least—worked to hide the fact that we were in the thick of it, in the here and now.[1] Waiting for culture encapsulated the aspiration for universal knowledge and the recognition that it would never arrive.

The goal of this book has been to explore what is often taken to be the margins of biomedicine and the outside of humanitarianism: laboratory animals. I have sought to argue that this sense of marginality is an error. Laboratory animals are embedded in almost every aspect of biomedicine and have also been actors in fomenting humanitarian thought and action in Britain. Medical treatment is central to humanitarian action today, and medicine relies upon making laboratory animals suffer and die. Waiting for culture, as a heuristic, helps to surface these muddles. I cannot provide a way out; I have no easy solution to the vastness of sacrificial logics that both humanitarianism and more-than-human humanitarianism rely upon. However, making visible the work involved in bioscience and biomedicine can show the extent to which human health depends upon the suffering and death of animals, and foster reflection on the inequalities that this instantiates—within and across species.

Britishness

The research that has informed this book has been rooted in the use of laboratory animals in Britain, which could be seen as a problematic form of methodological nationalism (Wimmer and Schiller 2002). The suggestion is that the nation-state acts as a container for social action and process. I would respond that, in some very important ways, the nation-state is indeed a container for scientific activity through funding, regulation, and institutions (e.g., Fourcade 2009; Yair 2019). But at the same time, science is very much a transnational activity contained by cosmopolitan concerns, which are also institutionalized (Mason 2016; S. R. Davies 2021; Tsing 2015). Therefore, how this study is contained in the United Kingdom is a necessary problematic that has ramifications for the concept of more-than-human humanitarianism.

It is a stereotype that Britain is a nation of animal lovers, and this is a stereotype that should be approached with caution. The unit of nationality and national cultures can give rise to the worst of what Karine Chemla and Evelyn Fox Keller (2017) refer to as culturalism in studies of science: fixed and essentialist units of an external society that are seen to determine the internal functioning of science (see Friese, Holmes, and Message 2023). Reuben Message (2023) has explored the stereotype of Britain as a country of animal lovers in this context, focusing on how this stereotype was weaponized in the Brexit campaign as an explicitly political tool. He refers to this stereotype as "animal welfare chauvinism," used not only by political elites trying to forge their version of Britishness but also by animal welfare advocates. The strategy was deployed across political positions (e.g., by the campaigns to Leave the European Union, to Remain in the EU, and by animal welfare activism), reproducing divergent imaginaries of a British community. As such, Message emphasizes that there is a decidedly top-down element to everyday life. He argues that, as researchers, we must listen carefully and critically to how these discourses play out in the everyday practices of animals, care, and science.

This book started with a statistical finding that reproduces the stereotype of Britain as a country of animal lovers and, thus, potentially, "animal welfare chauvinism." In the survey Nathalie Nuyts, Juan Pablo Pardo-Guerra, and I conducted of scientists—all of whom were working in the UK and according to the same regulatory requirements—we found that those scientists who identified as British were more likely to think that animal care was a crucial part of producing high-quality scientific research (Friese, Nuyts, and Pardo-Guerra 2019). We did not take this to mean that British scientists do, in fact, care more for laboratory animals than non-British scientists do. Such an interpretation would enact an attitudinal fallacy of interview research (Jerolmack and Khan 2014), or the mistaken belief that what people say and do is the same thing. Rather, we suggested that "care" and "animals" may represent a kind of taken-for-granted assumption, or civic epistemology (Prainsack 2006; Jasanoff 2005), that is linked to the history of humanitarianism in Britain.

In 19th-century Britain, animals provided an important reference for developing humanitarian thought and action. As a vulnerable group existing within the social milieu, animals—alongside the poor, the mad, slaves, women, children, the colonized, and foreigners (Haraway 1989; Thomas 1983; Ritvo 1987)—were deemed to require protection from tyranny and abuse. Treating animals humanely increasingly signified class status in the 18th century, especially across the 19th century. The need to *protect* those who are vulnerable encapsulated class, as it denoted the need for those who have the power to dominate to do so with responsibility. Learning to care for animals was thus a civilizing process (Tague 2015; Chakrabarti 2012). Protecting vulnerable others through benevolent paternalism became a way of maintaining privilege. More-than-human humanitarianism inherits, in an embodied manner, the ways in which care and hierarchy are entangled through protection that has shaped humanitarianism in Britain. In turn, it corroborates a particular way in which Britishness has been cultivated.

More-than-human humanitarianism is one way of asserting an idea of British identity, but there are, of course, other contesting ideals of Britishness. Not everyone in the 19th century worried about protection, and not all British people care about animals. Mette Svendsen (2023) has argued that a way out of methodological nationalism is not to forgo the nation as an analytic category; rather, we should forgo the understanding of the nation as an a priori and passive context in which science occurs. Svendsen prompts us to ask instead how versions of nation, nationality, and belonging are enacted in and through the connections and disconnections that are made in doing science. Who does the nation include? Which relations between people and animals as well as citizens and foreigners are promoted and which are restricted? In this conceptual space, it is possible to understand more-than-human humanitarianism as one way of asserting a British identity that occurs alongside other contradictory forms of Britishness. These forms may focus on forcefully removing people from the United Kingdom to other countries such as Rwanda, retreating from human rights in law and practice, and ultimately negating the imperative to protect others who are vulnerable.

To approach the nation as becoming, emergent through practices of both connection and disconnection, aligns with the theoretical framings of "partial connections" (Strathern 2004) and "being alongside" (Latimer 2013a). This book has sought to show how "being alongside" allows us to see how humans and animals are related and relating to one another, in practice and ideology, without collapsing key differences in the process. In this way, we can see how different ways of making Britain also connect and disconnect.

More-than-Human Humanitarianism

I have been adamant throughout this book that more-than-human humanitarianism is *not* a utopian concept. The dangers of how more-than-human humanitarianism risks being a form of animal welfare chauvinism make this abundantly clear. Describing the ideas and practices of more-than-human humanitarianism, this book instead seeks to make visible an ethos with a particular focus on affective practices and meanings. I do not see more-than-human humanitarianism as offering any kind of panacea for the ills of the world that we are living through today. There are serious limitations to the ethos of more-than-human humanitarianism, but this does not mean that there are not also benefits. There are reasons to worry about more-than-human humanitarianism slipping away. So, I will review and summarize those benefits here.

First, suffering is a social fact, and the shared focus on "alleviating suffering" for both humanitarianism and more-than-humanitarianism creates blind spots and sites of denial. I pause when confronted with the 19th-century antivivisectionists' belief that human suffering should not be ameliorated through animal suffering but rather through structural and individual changes in human behaviors (chapter 1). While I bristle at the religious moralism that undergirds this sentiment, which is rare today, I do think that it is an important reminder that alleviating the suffering of some through medicine and health care relies upon making another suffer and die. How might we alleviate suffering without making another suffer in our place is an important question that can be surfaced through more-than-human humanitarianism. Through this ethos we can see that the search for structural solutions to suffering, while necessary, should not be assumed to be free from causing other forms of knock-on suffering. Both Donna Haraway (2008) and Gail Davies (2012a) have argued that facing the fact of suffering is necessary to begin the work of amelioration. Elspeth did this: she faced a suffering rat as a fact and then tried to change the conditions that were making that rat suffer, not to save the rat but to ensure future rats would not suffer in this way. It is the imperfect work of more-than-human humanitarianism.

Second, care is a knowledge practice, and this knowledge requires embodied togetherness over time. I learned from animal technicians the empirical truth behind Eduardo Kohn's (2013) argument that *not* noticing a difference is a crucial component of knowledge generally and of knowing beyond the human specifically. Becoming habituated to a normal—so that its disruption can be registered as significant—is a crucial, if underappreciated, way of knowing, and it is how care as knowledge often works. There is no "waiting for care," and this is the lesson that Gifty learns in Yaa Gyasi's novel *Transcendent Kingdom* when she states: "I'm no longer interested in other worlds or spiritual planes. I've seen enough in a mouse to understand transcendence, holiness, redemption. In people, I've seen even more" (Gyasi 2020, 246). More-than-human humanitarianism is rooted in a relational conceptualization of mind (Kohn 2013), and an ethic of care (Gilligan 1982) as its way of knowing. It, therefore, allows us to bring subjugated knowledge practices to the fore.

Third, killing relations are a confronting social fact. Saviorism is both the exalted goal of humanitarianism while also perpetuating paternalism, racism, sexism, and colonialism. More-than-human humanitarianism is both problematic and useful because it is not rooted in saviorism but rather in ending suffering. This is important to bear in mind because "killing" those relations that are rooted in killing, as crucial as this is, will very likely be violent and will create suffering. A focus on saving human life, irrespective of the quality of that human life, can create new forms of suffering (Kaufmann 2006; Stevenson 2014). More-than-human humanitarianism usefully redirects attention to that suffering and asks how life can be made worth living.

Fourth, without a sacrificial logic, the sadness of necropolitics can become unbearable in a manner that can "sacrifice" those people who face all this death. More-than-human humanitarianism responds to the necropolitics that seems to underpin any biopolitical regime—here, the biopolitical regime of medicine. Witnessing suffering and death can take a tremendous toll. In this context, "sacrifice" is not only a moral economy (Svendsen and Koch 2013) but also an "affective economy" (Ahmed 2004). And here is the fix. Sacrifice is a universal, transcendental abstraction that sits uncomfortably within the Beckettian ethic of waiting itself. Sacrifice is Godot. How can attending to the suffering that necropolitics produces be approached without sacrifice as an affective economy? Who is left to do this attending (e.g., animal technicians, veterinarians)? How does this reproduce existing inequalities?

Fifth, compassion is a privileged *outcome* of the act of togetherness rather than a starting point for action. Compassion does not simply reside as an emotion within an individual. It is a socially produced experience and an accomplishment that challenges the idea of bounded, independent, and rationally self-realizing individuals. In this sense, more-than-human humanitarianism is aspirational in that it tries to evoke fellow feeling. By refocusing on the practices of compassion, more-than-human humanitarianism can extend what Arlie Hochschild (2016) has referred to as "empathy maps," widening out those who are deemed worthy of empathy and making the selectiveness of compassion a site of reckoning. It is through compassion that more-than-human humanitarianism overlaps with solidarity (Prainsack and Buyx 2017) and more-than-human solidarity (Rock and Degeling 2015).

Sixth, hearing and responding to a "Don't" is hard work and requires prelinguistic competencies. Hearing another say something that you or I may take as a "Don't" relies upon more than language; hearing a "Don't" requires intimate knowledge (Raffles 2002). How might we recognize the work that goes into hearing "Don't"? This work is necessary when working with other species, but it raises the question: Can people hear other people say "Don't" within relationships of power? How might this ability to hear and respond to a "Don't" be recognized and valued without trying to codify it? We learn from more-than-human humanitarianism the importance of bringing together multiple forms of care, not only skilled, haptic care but also regulatory, distanced care to address the dangers that not hearing another say "Don't" presents (see also Druglitrø 2018; Druglitrø and Asdal 2024; G. Davies et al. 2018).

There are many limitations of more-than-human humanitarianism. It inherits the paternalism and colonialism of its inception in the Victorian era, to be sure. It relies upon universal abstractions—truth, transcendence, sacrifice, grace—while undoing those abstractions in the process. It does not offer protection from being made killable. It enters the fix, but it does not provide a way out. I often think that people's best traits are also their biggest liabilities, which is true of concepts too. The benefits of more-than-human humanitarianism also contain its limitations. Suffering should not be a social fact. Not all knowledge is equal. Compassion makes for poor politics. Haptic care is overly romanticized. Killing is wrong. No one should be sacrificed. But here we are in the normative, whereas more-than-human humanitarianism tries to do the best one can in a world that is as others want it to be.[2] The aspirations of this concept are limited in responding to the world as it is encountered in the here and now.

Humanitarianism through the Prism

I see an analogy in the tensions between the figures of the human rights activist and the humanitarian aid worker, on the one hand, and the figures of the animal rights activist and the animal technician, on the other. Human rights and animal rights activists want to change the structural conditions that give rise to suffering, and the humanitarian aid worker and the animal technician respond to the fact that another is suffering before them. But where human rights activism and humanitarian aid have become increasingly blurred, the animal rights activist and the animal technician are in a polarized position to one another. The animal rights activist asserts that they care about laboratory animals through direct action and active resistance against their use in science. The animal technician asserts that they care by being with laboratory animals, caring for their needs and improving the well-being of laboratory mice—whom few others might be willing to care for with care.

What happens when we consider animal technicians' care as unique and important but also marginalized and threatened (see Greenhough and Roe 2019, 379–80)?[3] For me, the animal technician becomes the figuration of more-than-human humanitarianism. They sacrifice their lives in some very real ways to care for animals who will be born and die as part of bioscience today—as the story of Janet in chapter 5 shows, but also as the more mundane practice of taking on the burden of killing mice by dislocation of the neck to shift the stress from the mouse to the technician in chapter 4 shows. Didier Fassin (2007, 508) has argued that this freedom to sacrifice one's own life to work to save the lives of those deemed sacrifice-able—in his case, Iraqi people during the war on Iraq—asserts humanitarian ethics by reasserting the sacredness of a life that has been otherwise denied sacred status. Where humanitarians do this in the name of people deemed killable through war or disaster, more-than-human humanitarians do this in the name of animals considered as killable through their species-being.

Fassin is quick to point out that this principled ethic becomes troubled in practice, and he presents the stark differences in the life experiences of expatriate and national humanitarian aid workers to demonstrate the point.

> Distinctions are set up between foreign staff, almost always Western and white, and local employees. These distinctions, in addition to the material advantages conferred on foreign staff, are augmented by much more serious disadvantages that for the local staff, concern their very survival, whether they are endangered by illness or war. . . . Thus, within the humanitarian arena itself hierarchies of humanity are passively established but rarely identified for what they are—politics of life that at moments of crisis, result in the formation of two groups, those whose status protects their sacred character and those whom the institutions may sacrifice against their will. The protagonists in conflict are well aware of this distinction when they abduct people. They know that only foreigners have market value. Their compatriots are usually executed, as was the case in August 2006 for seventeen Sri Lankan Action contre la faim humanitarian workers killed by military forces. (Fassin 2007, 516)

While animal technicians are not at risk in the way that local aid workers are, they are far more likely to experience social isolation (Sharp 2019; Michael and Birke 1994; Birke, Arluke, and Michael 2007) and receive far lower levels of social rewards in the form of pay and recognition when compared to scientists. I have suggested that emphasizing the knowledge of marginalized people in these ways is one way to redress these inequities and challenge the status of these people—technicians or national workers—as sacrifice-able.

When focusing on knowledge, a key difference between humanitarianism and more-than-humanitarianism becomes clear. Humanitarianism is about saving lives; it is rooted in saviorism. More-than-human humanitarianism is not about saving lives; it is about ending suffering. More-than-human humanitarianism can provide a window into what happens when suffering is the focus. This focus is not without its dangers, to be sure. But what we see here of humanitarianism is that a focus on suffering can raise the question of how lives can be made, not simply to live, but to be worth living. And the opposite is also true. Through the prism of humanitarianism, it becomes possible to ask whether an animal has an interest in sustained life despite suffering.

Witnessing

The idea of a modest witness has been central to the development of science as we know it today, and science studies. Steven Shapin and Simon Schaffer's canonical book *Leviathan and the Air-Pump: Hobbes, Boyle, and the Experimental Life* (1985) showed how ideas about scientific truth and objectivity came to be located not only in a neutral instrument that sits at the center of the experiment but also through the confirmation of resulting findings by the aristocratic men assembled around the experiment. Technicians who made these experiments possible could not verify truth claims; they were not appropriate witnesses because of their class status. Donna Haraway (1997) added a gendered and racial analysis to Shapin and Schaffer's class analysis, showing how the gentleman scientist as a reliable witness worked to make both women and anyone who was not white marked, and also unable to verify truth claims. The idea of the gentleman scientist as the only reliable witness to a fact reproduced inequalities rooted in class, race, gender, and, as Haraway notes, was centrally concerned with Englishness. We still live with this history today: "The important practice of credible witnessing is still at stake" (Haraway 1997, 33). With the proliferation of fake news, it is probably even truer today than it was when Haraway was writing these words in 1997.

Witnessing is also a central logic of ethnographic methods. This book is rooted in the idea that it is important to see the work that goes into making medical knowledge and interventions that explicitly address human suffering. In turn, this requires facing animals who possibly suffer and will die for that knowledge. My research has been in close conversation with an interdisciplinary scholarship that has critiqued the "polarization cycle," wherein scientists and animal technicians are pitted against animal rights proponents such that objectivity and care are considered antithetical (G. Davies et al. 2024). To make invisible knowledge visible and to witness these otherwise invisible practices, I have had to traverse a space that is not neutral because of my emplacement within the spaces of more-than-human humanitarianism. This witnessing has thus been distanced from animal rights, by my having both *not* taken a political position and my entering the spaces of animal experimentation.

The dilemmas of this type of witnessing are not that it is disembodied, as the gentleman scientist proclaimed to be, but that it gains credibility through existing inequalities. These dilemmas also run across humanitarianism. Associated with Médecins Sans Frontières (MSF), who distinguish their approach from the neutral and discreet approach of the International Committee of the Red Cross (ICRC), the goal of witnessing in humanitarianism is to bear witness to and to report everyday practices that are otherwise invisible. Witnessing is a problematic practice here, too, for it is neither neutral nor does it take a political stance but rather speaks to the situated truths of what one sees.

In humanitarianism, this focus on witnessing and its links to biography is often viewed as a distinctly human affair. Didier Fassin states:

> It is rather that if one believes that what distinguishes humans from other living beings is language and meaning and that what makes human life unique is therefore that it can be recounted, as Hannah Arendt asserts, then humanitarian testimony establishes two forms of humanity and two sorts of life in the public space: there are those who can tell stories and those whose stories can be told only by others. With this new dividing line, life is no longer, as it was before, biological (the life that is risked or sacrificed); it is henceforth biographical (the life that is lived but that others narrate). (Fassin 2007, 518)

Much of ethology, animal studies, and science and technology studies challenge this dividing line between humans and other living beings because the dividing line—known as human exceptionalism—is an inequity that is interlinked with the second inequity that Fassin is interested in: those whose can tell stories and those whose stories can only be told by other people. In the context of laboratory animal science, social scientists and the occasional animal rights activist (who can enter an animal facility only under subterfuge) can stand as witnesses.[4]

Bearing witness disrupts discursive formations that seek to sanitize everything from war (e.g., as precise and humanitarian rather than violent, destructive, and painful) to science (as disembodied and, hence, neutral, objective, and universal rather than embodied, situated, and relational) and technology (as automating and thus freeing humans from degrading labor rather than socially constituted). But bearing witness tends to foreground the witness and witnessing is anxious about speaking for another. Givoni (2016) argues that witnessing is a technology of self across humanitarianism and ethnography that is centrally focused on creating an ethical self:

> I show that the analytical attention to the quandaries of humanitarianism is a medium through which Western physicians as well as other experts, who have come to dominate the humanitarian scene, now turn themselves into a moral personae who are equipped with the technical skills required to respond adequately to crises but are not fully determined by them. . . . I argue that this new mode of humanitarian reasoning should prompt us to probe both the humanitarian care for distant victims with care for Western selves. (Givoni 2016, 174)

I raise these resonant discussions about witnessing in order to address a key question that I have frequently been asked: How do I know that the scientists and animal technicians whose work has moved me and that I have described throughout this book *really* cared? How can I know that they weren't simply putting on a performance for me, pretending to care for mice as a show for the visiting sociologist? On one level, I cannot know what another thinks; that is the challenge that working with animals brings to the fore (chapter 6). And yet I do trust—when in the physical, embodied presence of many people as they work with animals—that this quite simply cannot be a performance. People's work has moved me, and this movement has served me in seeking to understand the affects of more-than-humanitarianism; I experienced this as quite real.

Social Change and Inequity

In the final chapter of *After Nature: English Kinship in the Late Twentieth Century* ([1992]1995), Marilyn Strathern offers a theory of change that knowledge practices both track and participate in. The goal of unpacking taken-for-granted assumptions has been a central project in anthropology specifically, which is Strathern's concern, but it has also been the goal of much of the social sciences and humanities more generally. This project, to make the implicit explicit, manifests social change according to Strathern:

> To bring to consciousness the context or grounding for one's sense of the world could not be confined to recognising what was already explicit, nor to making the implicit explicit, though we might have thought we were doing both of those things. Its effect has been to make context or ground itself disappear.... Postmodern aesthetics and Thatcherism alike most interestingly pull out from under our feet the grounding or reason for these constructs.... The sense is that context itself has gone.... If nature has not disappeared, then, its grounding function has. It no longer provides a model or analogy for the very idea of context. (Strathern 2004, 195)

Strathern argues here that unpacking assumptions takes away the grounding elements of those assumptions. The unearthing of some assumptions gives way to new assumptions. Strathern argued that, in the 1990s, choice replaced context in this manner, and she was concerned that merographic thinking was being lost in the process. She hoped to salvage ways of thinking about the connections between wholes that remain parts. Her concern was that when everything is simply a choice, a flattening occurs such that what is being reproduced is not bodies, ideas, or relations anymore but merely a choice in and of itself.

More-than-human humanitarianism marks how humanitarianism in Britain arose as a merographic concept; it connected the human with the animal while differentiating the two. Is it the case that this idea and ideal is also slipping away, disappearing as made explicit? I noted in the introduction that I see this book as potentially a form of salvage sociology. Different from the colonial problematics of salvage anthropology and corresponding problems of coloniality (TallBear 2017), salvage sociology may nonetheless be a collection of some of the affects that make up more-than-human humanitarianism as a concept—just as the ethos is being transformed into something else. My ability to name "more-than-human humanitarianism" may denote the very fact that it is changing—it is no longer taken for granted. And if so, what might replace more-than-human humanitarianism with what kinds of ideas about the world?

In chapter 4, I noted that sacrifice is less of an emic category and more of an analytic one. I very rarely heard scientists, animal technicians, or veterinarians use the term "sacrifice," but I did hear a lot of discussion of "cultures of care." This organizational morality has developed in response to high-profile and tragic failures of care across both human health care and laboratory animal science in the UK (Nuyts and Friese 2023; Gorman and Davies 2020). Richard Gorman and Gail Davies (2020) state that "the 'culture of care' appears as the 'culture of no culture' fails to protect people and animals from harm." In other words, a culture of care appears when the beneficence of a universal and transcendent science or medicine cannot be trusted to care for the vulnerable beings who require care in spaces that seek to be outside of society and culture. There are diverse ideas about what constitutes a culture of care, but going above and beyond legal requirements in providing care for both the caregivers and care receivers is a key goal (Greenhough and Roe 2018). A culture of care is also an attempt to undo hierarchies and thus redress social inequalities rooted in class and gender through organizational practices. When I visited a pharmaceutical company, I was told that the scientists had to change the bedding for the mice whenever they opened a cage. The goal here was to make scientists aware of and participate in the work of animal technicians while also providing better care for and understanding of the mice by attending to all their needs when disrupting their lives. It was one way in which this pharmaceutical company attempted to instantiate a culture of care.

How do we think about inequality in this context? Mike Savage (2021) has argued that Bourdieu's notion of fields no longer helps us to understand how inequality works because the economic field has become so dominant. If the economic field replaces other fields, Strathern's concerns that merographic thought is being replaced by the totality of choice or capitalist economics become all the more prominent. A culture of care could be seen to enact this economic rationale similarly. Yes, a culture of care is important in seeking to trouble hierarchies and inequalities to address failures of care, and it makes visible care work that is otherwise invisible. However, my critique is that a culture of care, like more-than-human humanitarianism, has limitations. One of those limitations, to my mind, is that the purpose of a culture of care is not framed vis-à-vis a greater humanity but rather as the basis for a more efficient and effective workforce. One needs to pause and consider the possibility that people may not only be alienated by having their knowledge subjugated but may also be alienated from themselves as carers in this context.

Cultures of care are increasingly being understood as something that can be measured—as seen in the Culture of Care Barometer within the National Health Service (NHS)—and therein governed. Care becomes part of a social process that Marion Fourcade (2016) calls "ordinalization." Metrics are being used to rank an increasing number of areas of everyday life ranging from universities to sports teams, employers to nation-states. Savage (2021, 112) has argued that ordinalization is not only a hallmark of neoliberal policies that seek to incentivize improvement but also—and crucially for my argument here—marks the entropy of any field. Savage states that ranking replaces the passion and intensity that brings a field together. Instead, the field becomes dominated by powerful "top dogs" who have the resources to ensure their position at the top of the hierarchy. I suggest that the metricization of care in science through things like the Culture of Care Barometer denotes entropy in caring fields through the weight of capitalism.[5]

Killing animals persists in this milieu of a culture of care but without the religiosity, inwardness, and sense of universal transcendence that is associated with sacrifice. A culture of care is thus meant to replace the role of sacrifice as an affective economy. In the process the abstractions related to sacrifice and transcendence may be giving way to new forms of governance. There are many reasons to think this could lead to improvements in animal welfare and animal technicians' treatment, so I am not advocating against the culture of care project. But the question remains: What kind of affective economies are available to replace sacrifice in this context, which make the constant presence of death manageable in laboratory research involving animals? Without an affective economy that creates a sense of purpose, the people who have to face the necropolitics undergirding our current biopolitical regime of bioscience and biomedicine remain at risk of themselves being sacrificed.

ACKNOWLEDGMENTS

Many scientists and animal technicians participated in this study, as well as quite a few veterinarians and Named Animal Welfare Officers (NACWOs) and a small number of animal welfare proponents and even some animal rights activists. Participation ranged from: attending a focus group to pilot the survey; completing the survey; helping us to boost the random sample that the survey was distributed to in order to ensure industry representation; participating in a qualitative interview; to allowing me to observe the work in an animal facility, laboratory, or both on a short- or a long-term basis. Participation also included having informal conversations.

Anonymity is particularly important when conducting social science research generally but is particularly important in the context of research regarding laboratory animals in the UK. In this context, some people experience anonymity as a precondition for their physical safety. And so I have decided not to name any of these people as individuals. I do realize that some people would have been happily named here, and I hope you recognize yourself and my appreciation of you in these pages. Empirical, social science research is not possible without the willingness of people to participate, and I greatly appreciate the time that you took out of your busy schedule. I have decided to name no one in order to ensure the anonymity of everyone.

One reason I decided to focus this book on the ethnographic fieldnotes I wrote as part of participant observation was because I could ensure that I was not putting under my name someone else's research. I say this because this book is based upon a collaborative research project, and I am indebted to the various members of the research team that at different times included: Juan Pablo Pardo-Guerra, Tarquin Holmes, and Nathalie Nuyts. This research could not have been done without the expertise that you each brought, including survey research, regression analysis, social network analysis, and historical research. Nathalie learned multiple correspondence analysis for all of us in this context. You were each a pleasure to work with and I feel grateful to have been able to collaborate with you all; thank you.

I could not have done the ethnographic research without the help of Joanna Latimer. Joanna introduced me to the scientist who introduced us both to the Institute, facilitating our ability to do ethnographic research alongside one another. One of my greatest joys of my academic life to date has been researching alongside and with Joanna, whose creative thinking and enthusiastic energy are an inspiration. While I was originally introduced to Marilyn Strathern's work as a general course student in anthropology at University College London back in 1995, Joanna has really been my teacher in thinking through partial connections rather than synthesis. Thank you, Joanna, for helping me to see ethnographically.

I have been very fortunate to have received funding from the Wellcome Trust to support this research. The Society and Ethics strand of the Wellcome Trust was at that time led by Dan O'Connor, and offered a supportive research environment before there was even a name for it. I felt the Wellcome Trust not only provided me with funding, but also a community of support particularly through the work of Paul Woodgate. For example, through a Wellcome Trust event for new grant holders, I met Robbie Duchinsky who has been a friend and support system in learning to manage a large research project. The Animal Research Nexus, also funded by the Wellcome Trust, brought many of the people involved in laboratory animal research together in various ways and over time, allowing for intellectual cross-fertilization that was particularly productive. A big thank you to Gail Davies, Beth Greenhough, Pru Hobson-West, Robert G. W. Kirk, and Emma Roe. I can't begin to tell you how much I have learned through you and your wider research team, including: Alistair Anderson, Bentley Crudgington, Richard Gorman, Renelle McGlacken, Reuben Message, Dmitriy Myelnikov, Alexandra Palmer, Sara Peres, and Tess Skidmore.

I have often remarked how lucky I am to work in a substantive field filled with so many kind, generous, and creative people. I cannot name everyone here but I want to say a special thank you to some key people who have really shaped my thinking through our conversations at various venues and who have influenced this book directly through our engagements on it in a range of formats, including: Kristin Asdal, Vanessa Ashall, Mie Dam, Tone Druglitrø, Sophia Efstathiou, Eva Haifa Giraud, Amy Hinterberger, Nicole Nelson, Astrid Schrader, Lesley Sharp, Jörg Strübing, Mette Svendsen, and Stefan Timmermans.

This research project has been a part of almost my entire career within the Department of Sociology at the London School of Economics. This is an incredibly kind and generous department, and all of my colleagues have supported me in one way or another over the years. I have presented this research at department seminars, and have received more helpful comments than I can recognize here. I would like to take this moment to thank those colleagues who specifically commented on this book, discussing key ideas across its development, including: Suki Ali, Ayça Çubukçu, Monika Krause, Claire Moon, David Pinzur, Sara Salem, and Mike Savage. I have had the pleasure to work with many amazing PhD students during this time. Thank you to: Valentina Amorese, Susanna Finlay, Reuben Message, Maria Kramer, Yan Wang, Martha McCurdy, Dominika Partyga, David Kampmann, Danielle Watts, Will Kendall, Shanae Dryer, and Babette May. Special recognition goes to: Reuben Message, who coorganized with Tarquin Holmes and me the conference on "National Cultures of Animals, Care, and Science" and has been discussing this project with me throughout; Will Kendall, who read carefully the chapter on sacrifice and has helped me to see the links with labor; Maria Kramer, who has discussed the challenges with ethnographic research of this kind and has buoyed me when I felt the story was simply too hard to tell; and Danielle, who served as a life-saving copyeditor as I finalized this text and who has been an interlocutor in thinking about care, animals, and science. I also want to thank Alice Christiansen, who was a visiting scholar in the Department of Sociology and has become a key interlocutor and friend.

I have been fortunate to have been in a writing group throughout this research project, and am indebted to my group that has over time variously included: Anna-Maria Carusi, Hanna Kienzler, Claire Marris, and Barbara Prainsack.

I am incredibly lucky to have had the opportunity to work with Ilene Kalish on a second book project; thank you, Ilene, for all your support. Thank you also for commissioning two incredibly helpful reviews of an earlier version of this book.

I received a research fellowship from the Leverhulme Trust to write this book, which gave me a year of research leave to devote to completing the first draft. I absolutely needed the time to think and focus on this book at precisely this time. And so I am incredibly grateful to the Leverhulme Trust for this funding.

This was from 2020 to 2021, and so coincided with the COVID-19 pandemic. My dad died on February 6, 2021, of cardiac arrest, during the second lockdown in the UK. My relationship to this book changed after his death, and it became a place for me to focus and escape from my grief. I was very close to my dad and love him and miss him terribly; he was a complicated person who was full of contradictions but who has been crucial to making me confident enough, as a person, to do academic research. And so this book is dedicated to him.

My daughter Hazel requires a special thank you as well. I had given birth to you, Hazel, four weeks before I had the interview at the Wellcome Trust for the funding that made the research in this book possible. You too have had to make time for this book, during lockdown and since. Thank you. I've learned everything I know about love and care from you. And to the rest of my family and friends, I love you and you make my life: my mom Carole, my sister Sarah, my brother Matt, Alicia, Trent, August, Roman, Richard, Dante, Steph, Helen, Linda, Andy, and Julie.

Finally, I would like to thank Adele Clarke who died before I finished this book. Just as writing the first draft allowed me to focus my grief after my dad's death, finalizing this book has focused my grief after Adele's death. It goes without saying that Adele was a magnificent mentor, and there is now a hole in my life where Adele once was. She built up the confidence in me that I needed to become a sociologist and a feminist researcher. I miss her terribly. But I was lucky to be writing this book as Adele, Rachel Washburn, and I were editing the second edition of *Situational Analysis in Practice*. Adele, Rachel, and situational analysis are imprinted upon me and this book. And I am lucky to be able to continue work on questions of methodology and situational analysis with Rachel Washburn. Thank you, Rachel; you are among my dearest friends and my most trusted colleagues.

APPENDIX

On Methodology

Having taught Qualitative Social Research Methods to MSc students at the London School of Economics (LSE) for several years, I have come to appreciate the importance of discussing in depth the methodology used in the research that is reported in a monograph. I have been inspired by the ways ethnographers have been more transparent about research methods—without breaking up the flow of the narrative of the book. Further, I have come to realize that it is not always clear how a large, externally funded research project, which is not only multisited but also mixed-methods, was conducted. This book is based on material collected through both a pilot study as well as the Wellcome Trust–funded research. In this appendix, I focus on the methodology and provide a discussion of the research instruments that the research team developed and used.

THE PILOT STUDY

The pilot study started after a biomedical scientist who works at a British university approached me. She told me that she believed the marginalization of husbandry and care that I was witnessing in zoos was not unique. Rather, she thought the marginalization of care work was pervasive across life science research and may be creating barriers to translational medicine. To further explain, she invited me to her laboratory so that I could see how she was incorporating improved animal care within the experimental system she had created to study and produce drugs to treat cardiovascular dysfunction. I started to make semiregular visits to this scientist's laboratory. I also attended relevant conferences with her, and she introduced me to other people involved in improving animal welfare as part of science. See also Friese (2013).

Through this pilot research, I started to formulate my specific research questions in classic grounded-theory fashion of alternating among data collection, data analysis, and research question refinement (Charmaz 2014). I was struck by people consistently remarking that it was "the usual suspects" who were attending the meetings focused on animal welfare—people who already believed that care for animals is of crucial importance for good science. People would ask how these meetings could reach scientists who might not see animal care in this way and, moreover, how I as a researcher would gain access to absent scientists' values and beliefs.

The scientists, veterinarians, and other proponents of improving animal care as part of doing science with whom I spoke had what I would call a sociological imagination regarding their field and how to do research in that field. This became a key point of our conversations. I would begin by explaining the usual STS research strategy: to follow key actors or actants around, to see how they made care central to science, who and what they involved or enrolled in the process, and where they experienced resistances. But the scientists, veterinarians, and other proponents of improving animal well-being in scientific research were skeptical of this methodological approach. The problem, according to them, was *accessing those people who care less* about animal care. They are not an organized social world, actively and collectively resisting efforts to embed improved animal care in science. Their resistance occurs in their silences and inattention. They do follow the regulations regarding animal welfare, and are unlikely to publicly bemoan them, but their attention is elsewhere, on other things. Thus the "cynical scientist" (an in vivo code, articulated at one meeting to capture the absent scientists) follows the rules but does not seek to go beyond the rules. Further, many believed that there were likely sociological patterns in terms of who strongly believed that animal care is important for high-quality science, and acted accordingly, and those who believed this less strongly (see also Holmes and Friese 2023). The scientists then offered their own ideas about how caring about animals is socially patterned among their scientific colleagues.

Given these experiences, I decided to formulate scientists' comments into four hypotheses about the social patterning of concern about laboratory animal care among scientists, and to test those hypotheses using a survey. The hypotheses we formulated, based upon the pilot research, included:

1. Attitudes about animal care are correlated with a scientist's position in the "field" (Bourdieu 2006, 1987, 1984) of science.
2. Attitudes about animal care are correlated with gender: female scientists are more likely to care more about animal care than male scientists.
3. Attitudes about animal care are correlated with age: younger scientists are more likely to care more about animal care than older scientists.
4. Attitudes about animal care are correlated with nationality: British/UK scientists are more likely to care more about animal care than non-British scientists.

I also formulated the overarching research questions of this project in relation to the wider social science and humanities literatures in addition to the pilot study. Laboratory animals are *not* an unstudied area of social life; there is an extensive literature in both the social sciences and humanities on animals in science that I used in order to ask critical questions about this situation. First, much of the contemporary social science literature on laboratory animals has to date used qualitative research methods, creating a rich scholarship on embodied care. But there have also been calls to extend the methods used in studying laboratory animals specifically—and animals in society more generally—in order to address more widespread social processes that frame the conditions of animals in society (Johnson 2015). I therefore decided to take up this call by drawing upon the ways in which situational analysis has long used multiple kinds of data, and to extend this to include survey data. It was through both the pilot study and the wider literature that I came to ask, "Why and how much do scientists in the UK think that animal care is important to scientific knowledge production?" Rather than studying the unstudied, one of Anselm Strauss's key maxims for innovative research, I used different methods in order to examine a well-studied situation in new ways.

CARE AS SCIENCE, WELLCOME TRUST–FUNDED PROJECT

This pilot study was very much the basis for the application I made to the Wellcome Trust. This research project as a whole asked how much and why scientists in the UK think that animal care is important to scientific knowledge production, how "care as science" as a value is practiced, and where this idea comes from. These questions were answered using: (1) a survey; (2) qualitative interviews; (3) participant observation; and (4) historical research. Nathalie Nuyts, Juan Pablo Pardo-Guerra, and I worked together on the survey and subsequent qualitative interviews; Joanna Latimer and I worked "alongside" one another—separately but together (Latimer 2013a)—in doing participant observation; and Tarquin Holmes and I worked together on the historical analysis. The project was designed so that the different parts built upon one another. The research received approval from the LSE Research Ethics Committee.

We started the research with the survey, with the aim of testing the hypotheses about the social distribution of animal care among scientists working in Britain that were derived from the pilot research. The survey was designed and implemented to explore scientists' (n=230) attitudes about animal care, which we piloted through two focus groups. The survey addressed the following topics: sociodemographics, career and work characteristics, attitudes and beliefs about animal care, social networks, and general values. The survey took 15–20 minutes to complete.

We followed a random sample procedure in selecting the respondents for a survey from a database that Nathalie Nuyts constructed of UK-based authors who had published an article on biomedical research, which used animals, between January 1, 2011, and December 31, 2014. From this database of 49,164 unique authors, we created a random sample. Taking into account possible outfall due to the mobility of researchers and missing contact information, we took a random sample of 2,000, with the aim of getting a final sample of around 1,000 scientists. For each of the 2,000 selected researchers, we checked the contact information (email and address details) manually with online available information. As this was a labor-intensive process the random sample could not be enlarged. In total, 1,164 scientists were contacted in the last week of June 2015 with a request to participate in our research by completing an online survey. To optimize the response rate, the initial email was followed by email reminders and a paper version was distributed by post in early September 2015. The survey had a response rate of 37%.

Because of the way the initial database was constructed, some of these respondents were not actively using animals in experimental research. Furthermore, some respondents did not fully complete the survey. In total we received 172 valid and completed surveys. Because the sampling methods resulted in scientists in industry and government being underrepresented, this group was additionally targeted by snowball sampling, which resulted in an additional 58 usable surveys. The snowball sample had a significantly higher percentage of women (63.8%) than the random sample (41.5%, $t(227)=2.98$, $p=.003$), and the respondents were significantly younger (average age of 38 in the snowball sample and 43 in the random sample, $t(219)=3.11$, $p=.002$).

The composition of the total sample was the following: 47% of our respondents were women while 53% were men. The average respondent was 42 years old ($SE=11$). Sixty-nine percent of the respondents identified as British nationals. The sample underrepresented managers, senior managers, and full professors (our sample had 21.07% against an estimated 35.6% in the UK) and overrepresented lower-managerial and research scientists (61.76% in our sample against 42.3% in the UK) as well as lower-status positions, including PhD students and laboratory technicians (17.15% in our sample against 14.3% in the UK) (Royal Society 2014). For privacy reasons the Home Office does not provide statistical information on gender, age, or any other characteristics of license holders in the UK (i.e., individuals licensed to undertake animal research), and so we cannot judge the representativeness of our sample on these variables. We can however assess other important variables, which do indicate a fairly good representation of scientists in the UK working with animals in their research. We analyzed the survey using multiple correspondence analysis, regression analysis, and social network analysis (see Friese, Nuyts, and Pardo-Guerra 2019; Nuyts and Friese 2023).

All survey respondents were asked if they were willing to be contacted for a qualitative follow-up interview to which 59 agreed. We followed up with all 59 individuals, which resulted in 14 survey respondents agreeing to a qualitative follow-up interview. These interviews were conducted by myself, Nathalie Nuyts, or both of us. Three interviews were conducted in person and 11 by Skype. Initial interviews were open coded by both Nathalie and myself independently, along the lines of constructivist grounded theory coding (Charmaz 2014). These open codes were then consolidated into 29 codes, and the remaining interviews were coded accordingly. The codes were then read and reread in conjunction with survey findings.

The survey findings did not go as I anticipated. I had thought that attitudes about animal care would be related to one's position in the "field" (Bourdieu 1987, 1984) of science. In other words, concerning the hypotheses, I anticipated that the amount of time spent with the research animal, which is typically linked to the stage of one's career, would help explain variations in attitudes about animals. However, much to my surprise, the major finding of the survey research was that those scientists who identified as being British were more likely to report that animal care was extremely important for science. I thus began the survey expecting to find that factors internal to the situation of the life sciences would relate to patterns in attitudes about the importance of animal care. Instead I ended up with a survey showing that nationality matters most in terms of how people report on the significance of animal care in producing scientific knowledge. We used both the interviews and the wider literature to understand this finding, in a manner that was consistent with abductive analysis (Tavory and Timmermans 2014).

This finding called for rethinking the subsequent subprojects of our broader project. First, I had originally thought that I would select different laboratories in which to conduct participant observation based on the field analysis. Is care practiced differently in large versus small labs, for example, or in universities versus industry? However, because attitudes about animal care did not map onto the field, or according to these axes of differentiation, such a sampling method would not necessarily allow the survey data and the ethnographic data to speak to one another. I therefore decided instead to conduct ethnographic research in just one site. I visited other university laboratories and animal facilities as I was looking to find a site to conduct participant observation, and this included one pharmaceutical company.

In the end, I conducted further ethnographic research from 2015 to 2017 at a large life sciences research institute in the UK. This is an independent research institute, which gains much of its funding from research councils that are funded by the state. It is affiliated with a nearby and highly esteemed university in training PhD students, but it is independent of the university. Much of the research conducted at this institute is "basic" science with a focus on aging, epigenetics, and immunology among other areas of biomedical science. However, some researchers do collaborate with pharmaceutical and biotechnology companies in doing more "applied" research. The animal facility that is part of this research institute houses primarily mice but also rats, and is quite large as animal facilities go. One veterinarian told me that, at any given time, there were up to 30,000 mice in the facility. This facility services not only the research institute but also pharmaceutical and biotechnology companies by breeding, rearing, and caring for mice and rats. The facility maintains itself in part through these contracts with industry. I could not, however, gain entrée into this part of the facility. For further descriptions of this research, see Friese (2019).

I conducted this research alongside Joanna Latimer, and we gained entrée by attending several meetings at the Institute. Through these meetings, we introduced our research and discussed how we would undertake a study of how the Institute models aging in its research. Over the next year and a half, we conducted ethnographic research by shadowing animal technicians, laboratory heads, postdoctoral and postgraduate research scientists, and a veterinarian. Given the nature of the site, we could never become "part of the furniture" in the way that the ethnographer strives for. The animal facility is biosecure, and so entry was highly controlled. In this sense, we as researchers had a "shadow-like" presence and were never woven into the Institute (Friese 2024). I took notes while conducting participant observation in the laboratories. Notes based on time spent in the animal facility were taken afterward because, since it was a biosecure facility, I could not bring a notebook with me.

The historical component of the research was also a direct result of the survey findings. Rather than exploring translational medicine as a field in which animal care emerges, I decided that it was important to understand instead the historical milieu through which animal care, Britishness, and science have come to mean something specific to the UK. Tarquin Holmes did much of this research by analyzing the 1875 Royal Commission on Vivisection that led to the 1876 Cruelty to Animals Act (Holmes 2021; Holmes and Friese 2023, 2020).

MORE-THAN-HUMAN HUMANITARIANISM, THE LEVERHULME-FUNDED PROJECT

Situational analysis allowed me to hold together several very different data streams in a multisited and mixed-methods research project. It allowed each data stream to be conducted on its own terms, while also keeping the different data streams in conversation and related to one another. This was invaluable in building the research program iteratively such that data analysis in one stream then shaped data collection and analysis in another stream. Earlier research was also revisited in light of later findings.

In this sense, the research was not strictly inductive but rather abductive. Toggling back and forth between data collection and data analysis alongside the existing social sciences and humanities scholarship on laboratory animals has most assuredly allowed me to develop *and refine* good research questions and strategies. Combining this emerging knowledge with extant theories, I was able to make informed guesses or hypotheses about what I would expect the answers to be and thereby collect data that can support but also call into question my own presuppositions.

Through situational analysis, I came to ask the new question of this book, which received funding from a Leverhulme Research Fellowship. Drawing on each research strand, I asked: What does the laboratory animal in Britain look like through the lens of humanitarianism and vice versa? If a history of humanitarian ideas, rooted in a benevolent paternalism toward vulnerable others that has included animals since the 19th century, forms a kind of embodied common sense for the British (Friese, Nuyts, and Pardo-Guerra 2019), then what might we learn about humanitarianism in Britain when viewed through the lenses of laboratory animals? Put differently, what does the situation of humanitarianism in Britain look like through the "prism" (Svendsen et al. 2018) of the situation of laboratory animals in Britain? And what does the situation of laboratory animals in Britain look like through the lens of humanitarianism?

Here I sought to further develop the comparative affordances of situational analysis through using Mette Svendsen's (Svendsen et al. 2018) "prism methodology." Svendsen has developed an approach that juxtaposes related but different ethnographic sites (e.g., a laboratory using pigs as a model for preterm infants in the ICU) by asking what does site A (e.g., the laboratory) look like through the prism of site B (e.g., the ICU) and vice versa. I extend Svendsen's prism methodology from ethnographic sites to situations, asking what situation A (e.g., laboratory animals) looks like through the prism of another situation (e.g., humanitarianism). I used situational maps to do this analytical work.

It is because of situational analysis that this book does not explore specific social spaces—the animal facility, the laboratory—but rather focuses on relational and situational moments that I experienced in doing participant observation. In other words, it is through situational analysis and abductive analysis that this book came to be about the affects of a set of care practices that I have named more-than-human humanitarianism.

NOTES

INTRODUCTION

1. For a historical perspective on the role of animal models in gerontology, see Brad Bolman's (2018) analysis of the use of beagles at UC Davis in the context of radiation research. Bolman shows that because dogs fight, housing them alone was a problem then as well.
2. As Beth Greenhough and Emma Roe note (2019), these decisions are certainly about animal welfare but they are also about cost. The expense of laboratory mice is normally determined by the cage rather than by the individual mouse. It would be considered prohibitively expensive to keep mice alone in a cage for two to three years.
3. This points to the ways in which (white) males have long been the standard by which universal, biomedical knowledge is produced—not only in clinical trials involving humans but also in preclinical trials involving animals. Over the past 30 years, this universal has been challenged and reconfigured in clinical research through the "inclusion" paradigm (Epstein 2007). When and how this paradigm shift in clinical research has (and has not) filtered into preclinical research is worth exploring.
4. I did not intend to focus specifically upon mice and rats in conducting this research, but it is not surprising that these are the species that I tended to see. According to the Home Office, mice were used in 59% of the 1.51 million experiments using animals within the UK in 2022, and in 86% of the 1.25 million procedures carried out to create or breed genetically altered animals (Home Office 2023). On the history of how mice became standardized model organisms in biomedical science see Lowy and Gaudilliere (1998) and Rader (2004); on transgenic mice see Haraway (1997) and G. Davies (2012b, 2013 a,b,c); for a social and cultural history of mice in relation to humans see G. Carroll (2014).
5. While a different story, a large number of research animals also had to be killed during the COVID-19 epidemic because of the difficulty in caring for these animals during a pandemic in which isolation was the key strategy. For a discussion of doing science during the COVID-19 pandemic, see Melanie Jeske (2024).
6. There is a literature, however, that has made the animal model a vital nexus in exploring bioscience and biomedicine, with a focus on the translational process within science and technology studies that I am building upon (Shostak 2007; Ankeny et al. 2014; Nelson 2013; Creager 2002; G. Davies 2010; Lewis et al. 2013). And Anna-Maria Carusi has begun the work of mapping the rare moments of

articulation, and thus the prominent absences, regarding animals in public discussions on COVID-19.
7 I am drawing on the sociology of ignorance here, which argues that not knowing something is not a background failure but rather a social accomplishment (Rayner 2012; McGoey 2007).
8 There is a body of scholarship that asks what happens if we understand laboratory animals not as tools, and thus things (Birke, Arluke, and Michael 2007), but rather workers (Haraway 2008; J. L. Clark 2014a; Porcher 2017).
9 It is hard to put the number of experiments involving animals into the wider context of biomedical research generally. Because scientific research involving animals requires a license in the UK, it is possible to count the number of experiments that are conducted and the number of animals used. There is not a centralized system that tracks how many biomedical experiments are conducted that do not use animals.
10 For a full discussion of the survey sample procedure, see the appendix. See also Friese, Nuyts, and Pardo-Guerra (2019) and Nutys and Friese (2023).
11 In this book I present ethnographic vignettes from both the pilot study and the ethnographic fieldwork that was conducted as part of the Wellcome Trust–funded research. Rather than code these fieldnotes, I instead wrote the fieldnotes into stories or vignettes in order to bring the spaces and interactions to life in a manner that put the reader in my shoes, and that was reflective of my own role (Humphreys 2005).
12 Tarquin did the primary analysis of this document as a trained historian, which we then conceptually analyzed and synthesized with the existing literature together (Holmes and Friese 2020, 2023).
13 There is an established literature showing that the rise of humanitarianism in Europe was linked to the rise of capitalism, and that humanitarianism was linked to governing through a bipolitical regime (e.g., Haskell 1985; Forclaz 2015).
14 This is a theme that I have explored in collaboration as well (Friese, Nuyts, and Pardo-Guerra 2019; Holmes and Friese 2023).
15 Many of the books that provide a history of humanitarianism are structured by the world wars and then Biafra, Kosovo, Rwanda, and Afghanistan (Fassin 2012; Barnett 2011; Rieff 2002).
16 Alasdair Cochrane, for his part, has argued that there can be animal rights without liberation. He maintains that animal rights tends to see any animal use as a source of abuse that needs to be abolished. This focus tends to be based in a set of equivalences made between humans and animals.

"The assumption is that once it is acknowledged that animals possess certain rights, it must also be acknowledged that they possess the right not to be used for certain purposes. In this sense, then, animal rights are often considered to be analogous to human rights. For human rights do not demand that we stop beating our slaves, or that we regulate the ways women are trafficked more humanely; instead they demand that such forms of exploitation be abolished and victims liberated. Many animal ethicists believe that animal rights have similar implications" (Cochrane 2012, 3).

Like Cochrane, I am arguing against this staging of equivalences. In foregrounding relations, I align with his argument that rights can instead be located in the idea that another being has interests and those interests impose duties upon others (Cochrane 2012, 2). This foregrounds the importance of relations, not only institutionalized relations but also interactional.

17 With the idea of laboratory animal lives and deaths as being "staged managed," I am drawing on and thinking with the work of Mette Svendsen and Lene Koch (2013) as well as Lesley Sharp (2019).
18 The idea of partial connections that shows how humans and animals are both connected and differentiated in humanitarianism—an idea that I make explicit through the more-than-human clause—is what distinguishes this concept from One Humanitarianism (Barona, Campos, and Martin 2024). One Humanitarianism is instead rooted in ideas about hybridity and holism.
19 Latimer and Haraway share many convictions. Both do their conceptual work vis-à-vis working/companion animals, Haraway with dogs and Latimer with horses. Both are critical of the ways in which human exceptionalism, rooted in a dividing practice, has given way to horrific violence against humans and nonhuman animals alike in the name of the sovereign subject who can "animalize" "othered" people (Latimer 2013a). And to find a way out of human exceptionalism, both reject the turn to a Deleuzian "becoming animal" through the carnality and animality of the human. And this is because neither Haraway nor Latimer understands any kind—human or dog or horse—as a figure that stands prior to relating. Becoming with and being alongside therefore share many assumptions. Where they differ is in what gets enacted through relating.
20 Indeed Latimer's (2000, 2013b) research agenda as a whole has been marked by a focus on how and when attachments, detachments, and reattachments occur in clinical spaces.
21 There is a history of medical science experimenting on humans in ways that have relied upon and reproduced structures of inequality based on race, class, sex, and nation (L. Briggs 2002). This is often through the equation of "othered" human groups with animals (Rothfels 2008; Ritvo 1987).
22 See the appendix for a full discussion of the methodology of the project, within which the question of this book and the ethnographic vignettes presented are situated.
23 I have followed Atanasoski and Vora (2019) in making this move, who note that their work on the racialized and racializing discourses of robotics allowed them to explore critical race and ethnicity without putting the burden of that research on racialized people.

1. SUFFERING

1 Laboratory animals in science are used as model organisms, meaning that these animals stand in for another species (Friese and Latimer 2019; Friese and Clarke 2012). A significant sociological, historical, and philosophical scholarship on animal modeling has focused on the epistemic and organizational aspects of this modeling logic for biomedicine. Model organisms are usually thought of here as "epistemic objects" (Rheinberger 2010, 154), around which entire disciplines can be built and different disciplines can come together (e.g., Kohler 1994; Shostak 2007; Creager 2002). A key distinction made by these scholars is between models representing biological processes occurring in another target species (e.g., a mouse represents a human in medical research) versus models that stand for a more basic process occuring across a full range of species (e.g., a biological process in a mouse stands for a more general biological process in mammals) (Bolker 2009; Ankeny and Leonelli 2011). The representational credibility of both types of models is constantly in question, however, and risks being dismantled if claims based on one species are overgeneralized to another species (Nelson 2013). Species generality through the evolutionary conservation of biological processes is thus a key assumption in modeling practices (Logan 2001, 2002, 2005), but one that is always also a site of concern, question, and debate (Ankeny 2007; Dam and Svendsen 2018).

My assumption was that there was a problem with modeling in research with endangered animals, linked to the generality and standardization of model organisms like mice that contrasted with the lack of standardization when endangered species are concerned. Laboratory animal bodies (e.g., flies, mice, rats, pigs, etc.) have been "standardized" in order to better represent human bodies, particularly in the context of biomedical "translations" (Clause 1993; Dam and Svendsen 2018; Kirk 2010, 2008, 2012; Logan 1999, 2002; Rader 2004; Kohler 1994). Selective breeding and transgenics enables laboratory animal bodies to be imagined as homogeneous in an effort to stabilize knowledge (Rader 2004; Davies 2012b, 2013a,c). Laboratory animal bodies are also standardized through routinized animal care that focuses on uniform practices of housing, food, and handling (G. Davies 2013b, 2010; Kirk 2010, 2014, 2008, 2012, 2016; Druglitrø 2018; Dam and Svendsen 2018). However, Gail Davies (2012b, 2013a,b,c, 2012a, 2010, 2011), Nicole Nelson (2018), and Mie Dam and Mette Svendsen (2018) have all shown that standardization is often unobtainable, even when genetics are held constant, as the very different spaces in which humans and laboratory animals live often thwart translational efforts.

2 In 2007, Charis Thompson (2013) argued to a group of stem cell scientists who had gathered in Asilomar, California, that the ELSI (Ethical, Legal, and Social Issues) approach to the life sciences that developed with the Human Genome Project had to be replaced. The politics of conducting stem cell science made it clear that ethics, law, and society were not simply implicated in downstream applica-

tions of scientific research, but rather in the conduct of science itself. In order to grapple with this, Thompson suggested another British name for women to serve as an orienting acronym. She argued that ELSPETH would capture the ethical, legal, social, political, economic, theological, and historical aspects of stem cell science. Inspired by Thompson's intervention, I used Elspeth as a pseudonym in order to articulate the ways in which care is viewed as part of science within the laboratory that I describe in the article on the pilot research (Friese 2013). Rather than viewing animal welfare as an extrascientific and regulatory concern, Elspeth (the person and the concept) sees care as integral to the experiment itself. See also Thompson (2013).

3   Robert G. W. Kirk (2014) has shown how the idea of "stress" transformed the ethics and eventual regulation of laboratory animal use in British science, expanding the scope of how an animal may suffer beyond physical pain—including social and mental distress. Building upon ethological knowledge, stress in these studies included not only how the animals related to one another in the cage but also how they related to the scientists and animal technicians working with them. From the 1960s to the 1980s, the focus shifted from not only mitigating animal pain to also promoting animal *well-being* (Kirk 2014, 251). Kirk focuses on how the human is imbricated in the production of animal well-being in this context: "Stress made the physical and social environment determining factors of the physiological state of the laboratory animal under study. Furthermore, stress relocated the human subject within that environment, making the researcher integral to, controller of, and obligated to the laboratory animals' well-being" (Kirk 2010, 258; see also Nelson 2018, 115–16 especially; Chiapperino 2021).

4   French's analysis focuses on the fallout from the 1876 Cruelty to Animals Act and the growing polarization between supporters of science and antivivisectionists. This polarization has been variously interpreted as: a sociocultural division between a professionalizing and institutionally expanding medical science with extensive intellectual and political pretensions and a reactionary antivivisectionist movement fearful of the "cold, barren alienation of a future dominated by the imperatives of technique and expertise" (French 1975); a gender and class divide between the apparent callousness of the "smooth cool men of science" and the politics of compassion espoused by feminists, socialists, and other social reformers (Kean 1998); an emotional divide between a calculating scientific sympathy and an intuitionist common compassion (Boddice 2016); and a contestation over the boundaries of the human and corresponding modes of biopolitical governance (Murphy 2014; see Holmes and Friese 2020). Tarquin's research troubled this focus on polarization itself. We built upon Shmuely (2017) who, for example, explores how empathy was institutionalized following the Royal Commission and with the 1876 Cruelty to Animals Act through the tandem processes of professionalizing not only science but also government in Victorian Britain. She shows how science and law were here "co-produced" (Jasanoff 2005; Reardon 2001). Tarquin and I developed the critique of polarization, but instead concentrated on

the Royal Commission that preceded the 1876 Cruelty to Animals Act, and on the construction of the anesthetized animal as a "boundary object" (Star and Griesemer 1989) that allowed for "cooperation without consensus" (Star 1993). Complicating the traditional science–antivivisection dichotomy, we thus explored the ways in which scientists themselves were conflicted about vivisection. We further argued that the 1876 Cruelty to Animals Act represented as much a negotiation within the scientific community as it was between scientists and humanitarians (Holmes and Friese 2020).

5   Boundary objects (Star and Griesemer 1989; Star [1988]2015) are entities that exist at the junctures where different "social worlds" interact in an "arena" or an area of sustained and shared interest and concern. Tarquin and I (2020) have shown how the anaesthetized animal, as a boundary object, was a focus for the anxieties of scientists regarding not only the nature of pain but also the relationship between life and death and the limits to trust in science. Adele Clarke and Susan Leigh Star (2008, 113) have defined the social worlds framework as focusing "on meaning-making amongst groups of actors—collectivities of various sorts—and on collective action—people 'doing things together' (Becker, 1986) and working with shared objects, which in science and technology often include highly specialized tools and technologies (Clarke & Fujimura, 1992; Star & Ruhleder, 1996)." If social worlds are held together by a shared technology, and vivisection was increasingly becoming a crucial technology for holding physiology together as a social world at the end of the 19th century in Britain, Holmes and I have argued that scientists themselves were conflicted about their social world. Indeed, these types of conflicts are central to the social worlds concept, as Clarke and Star continue (2008, 11): "Over time, social worlds typically segment into multiple worlds, intersect with other worlds with which they share substantive/topical interests and commitments, and merge. If and when the number of social worlds becomes large and crisscrossed with conflicts . . . the whole is analysed as an arena. An arena, then, is composed of multiple worlds organized ecologically around issues of mutual concern and commitment to action." Tarquin and I have argued that combining the technology of anaesthesia with vivisection allowed for scientists to persist as a social world, to continue to be held together as opposed to splitting. This required separating out "immoral" scientists as a distinct "subworld" or "segment" composed of "mavericks" in order to maintain legitimization (Clarke and Star 2008, 118–19). Emanuel Klein embodied such a maverick. French states that "without the testimony of Emanuel Klein, several members of the Commission would have been entirely unwilling to sign a report recommending legislation of any kind" (1975, 103).

6   I say this based on conversations I have had with social scientists working in Europe. These conversations often highlighted that while we were witnessing similar ethical practices and procedures, the affects we encountered in our research appear to have been distinct. That said, further research would be required in order to confirm this. I note that I simply say that these affects may not arise; they may as well.

7   Following mice, rats are the most commonly used species in laboratory science. In 2022, rats accounted for 12% of the 1.51 million procedures carried out in the UK for experimental purposes. For a historical analysis of rats as model organisms see Clause (1993) and Logan (2001, 2005); for a social and cultural analysis of rats in relation to humans see Beumer (2014) and Burt (2006).

2. CARE

1   There is a large and theoretically diverse literature on care. While this chapter is framed by tensions between political–economic and knowledge-based approaches, by and large my research approach to the theme of care in science has been informed by science and technology studies (STS). This literature has focused on care as a practice, which means that—as a practice—care can be both a knowledge practice and a site that is shaped by political and economic factors like neoliberalism (Latimer 2000; Atkinson-Graham et al. 2015; Atkinson, Lawson, and Wiles 2011; Ducey 2010). Much of the STS literature on care goes through the work of Annemarie Mol (2008) and Maria Puig de la Bellacasa (2011, 2015, 2017). Mol (2008) importantly intervened in the idea that choice is the only response to the problem of paternalism in the specific context of health care by presenting the logic of care as an alternative (Mol 2008). Her concern is that "choice" hides neglect, and this concern has been empirically validated (Pilnick 2022). Mol and her colleagues (Mol, Moser, and Pols 2010) emphasize that in this context care must be understood as a practice, one that cannot be judged as "good" or "bad" in universal terms. This critique of the normative assumptions that are often linked with care was further developed in a special issue of *Social Studies of Science* (Martin, Myers, and Viseu 2015). The ways in which care is also a site of control have thus been an important site of investigation within STS, and from the perspective of practices (Singleton 2010; Murphy 2015). In this context, technology and care are not antithetical; rather, technologies can be made caring, or not (Mol, Moser, and Pols 2010; Mol 2008; Pols 2010; Oudshoorn 2011; Boris 2010).

While I tend to draw upon this more descriptive approach to care, I am also informed by the feminist STS approach that, while recognizing how care can cement hierarchies, nonetheless holds out the possibility that foregrounding care is the basis for doing science better (Haraway 2008; Puig de la Bellacasa 2011). One of the key formulations of care in STS is offered by Maria Puig de la Bellacasa (2011), who addresses this question directly. Puig de la Bellacasa calls for an ethos of care within social studies of science and technology. With "matters of care," Puig de la Bellacasa wants to direct analytic attention to care as an affective state, a material way of doing, as well as an ethicopolitical obligation that includes the researcher's own sense of responsibility. In other words, care is at once an emotion, a practice, and a politics. Puig de la Bellacasa emphasizes that care is a neglected and yet highly constitutive world-making practice in terms of both the material world and the ways we come to know it. Building upon the work of Bruno Latour as well as Isabelle Stengers, the focus for Puig de la Bellacasa is on creating connections or assemblies through the work of science studies itself. See also Friese (2016) for this discussion of Puig de la Bellacasa in relation to veterinary medicine.

2 Gail Davies (2012a, 3) has argued that "no enrichment results in unequivocal welfare gains for the two different mouse stains" that she is studying, which in her research included ICR(CD-1) mice and C57Black 6 mice. Importantly, Janet is thinking about not only a stain-specific enrichment but also an age-specific enrichment for these BALB/c mice.

3 My Marxist feminist approach to care and reproduction is influenced by feminist technoscience, and particularly the work of Donna Haraway (1997), Sarah Franklin (2007, 2013), Judy Wajcman (2004), and Anna Tsing (2015). This work tends to avoid the term "social reproduction," as part of a critique of the idea that social reproduction and biological reproduction can be clearly delineated. But I am aware of and interested in Marxist feminism that does focus on social reproduction, and critiques its erasure under Marxism historically and today. Wages for Housework importantly emphasizes that care work is unpaid or underpaid, and this is gendered and racialized. See, for example, Silvia Federici (1975) historically and Mai Taha (2023) or the Care Collective (Collective et al. 2020) today.

4 Isabel Briz Hernández (2024) refers to the intersubjective practices that are required for translational medicine to conduct "miscellaneous care." She argues that intersubjective skills are capitalized alongside biological process and substance in this context.

5 In her history of the international move to standardize laboratory animals in the mid-20th century, Tone Druglitrø (2018) shows how this included not only genetic standardization through breeding (Rader 2004) and infrastructural standardization through housing (Kirk 2016; Druglitrø 2016; Bjorkdahl and Druglitrø 2016) but also standardization in caring practices (G. Davies 2012a). This included transforming the workforce of animal caretakers into animal technicians, who would "see skillfully" (Druglitrø 2018, 658). Beth Greenhough and Emma Roe (2019) cite "seeing skilfully" as a key element of animal technologists' expertise today, which allows them to acknowledge laboratory animals as embodied and lively subjects. They also note that this is a skill that ethnographers struggle to acquire (Greenhough and Roe 2019, 372).

6 On the theme of heat, comfort, and sleeping, see also K. Thompson and Smith (2014) who have analyzed the literature on people and dogs cosleeping.

7 Heat cannot be understood as a "mouse word" in the way that Mariam Motamedi Fraser (2019) describes "dog words," which I discuss in chapter 6. The mice are not using their bodies in order to intimate something; rather, the warmth of their bodies is taken as significant by Martine, and she acts accordingly. This follows Peirce's model of communication, which Kohn extends beyond humans.

8 Whether intended or not by Gyasi, the story of Gifty seems to respond to 19th-century antivivisectionist Frances Cobbe's charge that science is an "overestimate of Knowledge as compared to Love" (in French 1975, 7).

9   Where women represented only 8%–10% of the profession in the United States in 1970, by 2008 the gender ratio was almost equal (Irvine and Vermilya 2010, 58; Narver 2007, 1798). And women are likely to represent a majority of practitioners within the field in the near future. In 2008, 79% of the applicants to US veterinary-medicine colleges were women and 20% were men (Irvine and Vermilya 2010, 58). An increasing number of women within veterinary medicine is also the case in Canada and many European countries (Irvine and Vermilya 2010, 58; Koolmees 2000; Lofstedt 2003). In her study *Valuing Animals: Veterinarians and Their Patients in Modern America*, the American historian of medicine Susan Jones (2003) provided a historical overview of the development of veterinary medicine in the United States across the 20th century. She shows that veterinary medicine professionalized at the beginning of the 20th century with a focus on agricultural animals. Prior to this, "animal doctoring" was associated with the stable and the barnyard, and thus with the work of men of a lower social class (2003, 11). Jones shows how this image formed a barrier to the professionalization of veterinary medicine, while also marking it out as an extremely male-dominated profession (2003, 12–14). For example, veterinary medicine was considered far more masculine than human medicine because there was not a caring component in the former but there was in the latter. Jones shows that veterinary medicine at the turn of the 20th century demonstrated a masculine ethos by emphasizing (1) economically valuable animals over companion animals and (2) force, brutality, and an unsentimental approach to animals over care, compassion, and an emotional awareness of animals. As such, veterinary medicine developed as a clear "masculine profession" (Irvine and Vermilya 2010).

Jones goes on to show that veterinary medicine came to focus increasingly on companion animals in the 1920s, and by the mid-20th century this was the most stable patient population for the profession (Jones 2003, 116, 122). The shift from large, economically valuable animals to small, sentimentally valuable animals required departing from the more masculine tropes of veterinary medicine. Veterinarians knew as early as the 1920s that they would have to conduct themselves differently with the (female) owners of small companion animals than with the (male) owners of large agricultural and working animals (2003, 123). From the 1930s on, animal welfare became part of how veterinarians characterized their mission and an emotional value for animals became increasingly accepted (2003, 116).

Jones notes that the shift from large- to small-animal practices did not have an immediate effect in terms of increasing the number of women in the profession (Jones 2003, 139–40). While having women in a practice was viewed as useful or important, these women were generally the receptionist, bookkeeper, or veterinarian's wife (2003, 139). However, in the 1970s, US veterinary colleges were forced to admit more women through Title IX of the US Education Amendments of 1972 (Jones 2003; Lincoln 2010), which states that no person can be excluded on the basis of their sex from participating in or benefiting from any educational program or activity that receives federal financial assistance. An increasing number of women within veterinary-medicine colleges, combined with wage stagnation relative to law and medicine, has resulted in a declining number of applications to veterinary-medicine colleges by men (Lincoln 2010). In this context of professional feminization, women have been seen as particularly well-suited for work with companion animals. Indeed, numerically, women dominate the more lucrative companion-animal industry, while men dominate the less lucrative food-animal industry (Lincoln 2010). See also Friese (2016).

10 Drawing on Gilligan, a 2007 paper titled "Demographics, Moral Orientation, and Veterinary Shortages in Food Animal and Laboratory Animal Medicine" by veterinarian Heather Lyons Narver (2007) argues that women are more likely to exhibit an orientation toward relations of care while men are more likely to exhibit an orientation toward universal notions of justice. She argues that this means the influx of women into veterinary medicine will change the profession. Indeed, Narver understands differences in moral reasoning to explain, in part, why women veterinarians disproportionately work within the companion-animal industry while men disproportionately work in food-animal and laboratory animal medicine. Where working within the specific circumstances of companion animals and their human owners often requires a relational and situated approach to moral reasoning, a universalizing approach is more likely to be well accepted by scientists and agriculturalists. Further, Narver contends that an ethic of care is becoming dominant in the companion-animal sector, exemplified by bond-centered care. For example, women veterinarians have been found to practice relationship-centered appointments with clients more often than men in companion-animal practices; this includes talking more with clients, providing more positive and rapport-building comments, and being perceived as less hurried, which results in clients who provide more information to women veterinarians compared to men veterinarians (Shaw et al. 2012). However, Narver argues that an ethic of care is particularly needed within food-animal and laboratory animal medicine, both of which remain male dominated. In both contexts, Narver (2007, 1803) believes that an influx of women veterinarians could result in animal use being scrutinized more closely and critically because of a greater reliance upon an ethic of care, and that this would ultimately improve the living conditions of the animals.

Meanwhile, the sociologists Leslie Irvine and Jenny R. Vermilya have argued in "Gender Work in a Feminized Profession: The Case of Veterinary Medicine" (2010) that, while veterinary medicine is becoming a woman-dominated profession, the gendering of the field remains masculine in terms of the expectations of workers and their attitudes, behaviors, and interactions—and this includes small, companion-animal practices. For example, although the women veterinarians they interviewed did express a belief that veterinary medicine is a good profession for women as a caring profession, almost all the women they spoke with went on to distinguish themselves by saying that they were personally drawn to the field because they enjoy science. Further, the women described having to adopt stereotypical masculine traits, and dis-identify with stereotypical feminine traits, in order to be viewed as a professional. Irvine and Vermilya state:

> "Although women are not overtly asked to 'keep their place,' the attributes considered feminine are devalued even—and perhaps especially—by women, who risk devaluation by association. This is exemplified by women's eagerness to distance themselves from the caring, nurturing side of veterinary medicine and to emphasize their more clinical interests. Attributes associated with the female are disregarded in favor of those considered more masculine, lest the possessor of those attributes be considered unprofessional. In this way, an occupation that has feminized in numerical terms can remain masculine in other ways. The masculine culture is reproduced and maintained by those whose interests are at odds with it." (Irvine and Vermilya 2010, n76).

Irvine and Vermilya's analysis therefore troubles the idea that an increasing number of women in veterinary medicine will necessarily lead to ideological and practical changes in matters of care for either other women within the profession or for animals. They dissent from both Gilligan and those whose arguments have cited Gilligan, as they do not find evidence that an increasing number of women will change the ideology and practices of veterinary medicine. See also Friese (2016).

11 This argument is inspired by Anna Tsing's (2015) claim that an important connection between the economy and the environment lies in the history of the human concentration of wealth through making both humans and nonhumans into resources for investment. "This history has inspired investors to imbue both people and things with alienation, that is, the ability to stand alone, as if the entanglements of living did not matter. . . . This is quite different from merely using others as part of the life world—for example, in eating and being eaten. In that case, multispecies living spaces remain in place. Alienation obviates living-space entanglement. The dream of alienation inspires landscape modification in which only one stand-alone asset matters; everything else becomes weeds or waste." (Tsing 2015, 5–6) Alienation occurs in the animal facility not simply because people sell their caring labor for a wage, but because the knowledge produced through care is erased in the name of one science alone that is scalable as both universal and commercial.

3. KILLING

1 For a cautionary warning about the dangers of careful killing when used with humans, see Susan Benedict's (2003) analysis of nursing in Nazi Germany; in warning against euthanasia to relieve human suffering, she states:

> The first victims of the euthanasia program were handicapped children. In 1938, the father of a severely handicapped child appealed to Hitler for permission to have the child killed. Hitler instructed his personal physician, Dr. Karl

Brandt, to examine the child and if he found the child to be as handicapped as described, he was to have the child killed. The child was killed and the killing of other mentally and/or physically handicapped children began.... When individuals and organizations contemplate positions on today's debate over assisted suicide and euthanasia, it would be valuable to at least be aware of a period in the not-too-distant past when there was a rapid slide down the slippery slope to the killing of vulnerable populations. (Benedict 2003, 62)

2 See also Brad Bolman (2018) on the necropolitics of laboratory research involving animals.

3 This can be seen as an example of the "sociozoological scale" (Holmes 2021), as well as "the sentimental structure of laboratory life" (Sharp 2019). It is worth noting that laboratory animals are not at the bottom of this scale, however. Noemie Merleau-Ponty (2019) shows how developmental biologists create a hierarchy with embryos having less value than animals who have less value than humans in arbitrating the meaning of deaths in laboratory science. Joanna Latimer and I found that the deaths of mice were of far more concern than worms (Friese and Latimer 2019).

4 For more on the number of mice used in research in the UK, see Hannah Hobson, "Animal Research Statistics for Great Britain, 2018," *Understanding Animal Research*, www.understandinganimalresearch.org.uk, July 18, 2019 (accessed May 5, 2023).

5 Further, transgenic animals are not considered safe for rehoming, despite resistance to this in more charismatic species like pigs (J. L. Clark 2014b).

6 In the context of foot-and-mouth disease in the UK, John Law (2010) traced the practices of killing by veterinarians. He found that killing was a practice they used in relation to four different objects of care: for the animal, for the farmer, for the self, and for various versions of the collectivity. Law emphasizes that even this list is a reduction, as the objects of care were even more multiple. How to "choreograph" (C. Thompson 2005) this multiplicity is the work of veterinary medicine, according to Law:

"Annemarie Mol talks of the importance of tinkering in medical care. She treats the latter as a set of constantly unfolding and only partially routinised practices for holding together that which does not necessarily hold together. And this is the nature of veterinary care too: it can be understood as an improvised and experimental choreography for holding together and holding apart different and relatively non-coherent versions of care, their objects, and their subjectivities. It is the art of holding all those versions of care in the air without letting them collapse into collision." (Law 2010, 69)

This understanding of killing and caring, and of the art of veterinary medicine, aligns with the focus on partial connections and being alongside that frames the analysis presented in this book.

7 Weil (2006) maintains that clear, responsible action toward a nonhuman animal, as killable, can entail either killing or not killing. Weil traces this affective re-

sponse, which requires an "everyday morality" (Sharp 2019) as opposed to a codified ethics, through a "counter-linguistic turn" that has cut across animal studies, literary studies, and disability studies (see chapter 6). This counterlinguistic turn understands language as an obstacle to knowing, thus reevaluating the Heideggerian idea that only humans experience "death" as such by way of language. This is encouraging according to Weil (Weil 2006, 89). Nonetheless, she is concerned that, while the fact of a human person killing an animal has been used to raise the limits of language and can even take on something of a posthumanist religious experience for the human in question, the fact of killability persists. "What remains unaffected is the sacrificial structure that violently re-establishes those boundaries at the moment they appear to be effaced. It is, of course, the animal alone who dies or at least perishes" (Weil 2006, 92). I believe that Weil is urging those of us interested in the everyday moral practices of the laboratory to keep competing interests in the frame: might the animal have an interest in future life?

8 In her literary analysis of *Disgrace* and its posthumanist ethics, Calina Ciobanu (2012, 668–69) traces this statement to Thomas Hardy's novel *Jude the Obscure*. Where human children were, tragically, *too menny* in Hardy's novel, the dogs are too in Coetzee's novel. In Hardy's novel, "too many" is intentionally misspelled "too menny" in order to highlight the overwhelming desperation of the working class in Victorian era England. I will consider Ciobanu's analysis of Coetzee's posthumanist ethics further in chapter 6.

9 See also Sharp (2019, 45) for a discussion of infanticide among laboratory mice, a practice that the animal technician she is speaking with locates in animal care practices—specifically the problem of oversurveillance.

10 Rodante van der Waal (2024) has explored how accusations of infanticide were prominent in the early modern witch hunts, and argues that obstetrics enacts a form of accusation that inherits this history such that when women do not comply with medical advice they are at risk of being labeled a "baby killer."

11 For example, the Royal Society for the Prevention of Cruelty to Animals (RSPCA) did investigate if the lawyer Jolyon Maugham should be tried for killing a fox with a baseball bat after the fox got trapped trying to get into a henhouse. The RSPCA reportedly did not pursue this as a criminal case because the fox could not be proved to have suffered unnecessarily (Mohdin 2020).

4. SACRIFICE

1 This can be seen as a more general diagnosis of Victorian-era Britain, one that Bentham helped to articulate. For example, Tarquin Holmes and I (Holmes and Friese 2020) cite Samuel Haughton, who opposed vivisection for scientific demonstrations and suggested to the 1875 Royal Commission that other universities should do what Trinity College Dublin, his own institution, had done and use "animals freshly killed, in which you can keep up artificial respiration" (Royal Commission on the Practice of Subjecting Live Animals to Experiments for Scientific Purposes 1876, 103). "Unlike cruelty, killing animals was broadly

accepted in Victorian British society so long as it was for a suitable purpose—e.g., food, pest control or socially acceptable sport—and not unnecessarily painful. It was also viewed as a preferable alternative to even quite minor animal suffering" (Holmes and Friese 2020, 50).

2 I am not alone in witnessing and experiencing death as entirely unremarkable. See also Mette Svendsen (2022) and Eduardo Kohn (2013) on this theme.

3 See also Sharp (2019) for a discussion on the debates over reusing laboratory animals. For a telling counternarrative, see Richard Gorman's (2024) analysis of horseshoe crabs whose blood is used to test the safety of vaccines, injectable medicines, and medical devices for human and veterinary medicine. Gorman analyzes how this animal falls outside of the category of laboratory animals, and the protections this category bestows. As a "wild" animal, there is little concern about the impact of capturing and bleeding horseshoe crabs. Release works to justify the practice, even if this means an individual may be reused.

4 Svendsen (2022, 62) notes this affective element in sacrifice as well, in her quote of one scientist who said that working in connection with clinical medicine made her work with laboratory animals far easier than if she were to, in contrast, do laboratory research for the cosmetic industry. Here too we see how using animals and killing animals is easier when there is a sense of purpose.

5 Daston's (1995, 6) delineation of moral economy as a process through which scientific knowledge practices and exchange relations are informally regulated thus differs from Mertonian norms, as moral economies are understood as historically and culturally created and thus subject to change and contestation over time. This is where we can see Daston's moral economy as compatible with E. P. Thompson's (1971) delineation of "moral economy" through his argument that the food riots in 18th-century England cannot be explained in terms of economic reductions (e.g., unemployment, hunger, and distress). Rather, the food riots were a response to the changing social organization of rights and customs. Specifically, the paternalist model rooted in provision, upon which the poor depended to ensure the price of bread in times of dearth, worked to legitimize the actions of the crowd. The food riots thus expressed the moral economy of the poor. The incursion of a "free" market rationality was resisted by imposing prices through crowd action within the marketplace (E. P. Thompson 1971, 117). "Moral economies" thus denotes a set of historically specific expectations about the morality of social relations, which enable and legitimize certain forms of action and protest when incurred. A number of scholars have used these differing and yet compatible concepts of moral economies in order to explore laboratory animals (Kirk 2016; Koch and Svendsen 2015; Svendsen et al. 2017; Svendsen and Koch 2013; Sharp 2019).

6 Svendsen's analysis aligns with what Eduardo Kohn (2013, 17) calls "the general problem of how death is intrinsic to life." What Kohn sees, as Runa kill animals as food in the forest, is that hunting, fishing, and trapping all require that people assume the points of view of animals in a manner that understands those animals as also selves. What I think Svendsen and her colleagues show is precisely how

this more general problem of how death is intrinsic to life gets played out in laboratory work. People in labs have to take up the point of view of the individual mouse or pig to ask if the animal is suffering too much to be called upon to continue to live for a science that seeks to better the lives of another set of species. Because death is intrinsic to life, Kohn (2013, 17) argues that objectification is not the opposite of animism but rather the flipside; the act of killing always comes with the understanding that one can also be killed. Drawing on Kohn (2013, 17), the importance of death for life itself in the laboratory can be both an overwhelming contradiction, creating a "feeling of disjuncture," while also being "completely unremarkable."

7 There is a tension between protecting individual animals and populations of animals. One of my interlocutors pointed out to me that this can be seen in the differences between the focus of the Royal Society for the Protection of Animals (RSPCA) on protecting individual animals from harm when compared with the focus of the Royal Society for the Protection of Birds (RSPB) on protecting populations of bird species.

8 Using ethnography to chart how the Severe Acute Respiratory Syndrome (SARS) outbreak in 2003 facilitated an ongoing shift in China's public health system, Mason (2016) explores the consequences of replacing Mao Zedong's famously low-technological and deprofessionalized approach to health care that was rooted in the barefoot doctors working at public health posts (and that had been credited with a number of improvements in health, such as life expectancy and infant mortality—both of which are used as barometers of a nation's health status). These crumbling public health posts were replaced with Centers for Disease Control and Prevention during China's transition in economic development following Mao's death, and were named specifically after the CDC in the United States. Mason argues that this entailed a shift in conceptualizing the commons, "geared toward the protection of global, rather than local, interests and toward the protection of a cosmopolitan middle-class dream rather than toward the betterment of the poor" (Mason 2016, 3). This shift was made logical in the context of global science and public health, in that practitioners learned to govern local populations on the basis of an idealized notion of modernity, science, and professional trust. But the asymmetries of global public health were made apparent as its project was pursued. For example, while China's use of quarantine was lauded to control SARS within its population, China was condemned when quarantine was used with North Americans traveling to China in order to stop H1N1 from entering its population.

5. COMPASSION

1 For a discussion of animal ethics and compassion that is rooted in the body, see Ralph R. Acampora (2006). He notes that animal rights is stymied because: "Despite the best efforts of many animal ethologists and ethicists . . . there persists—at least amongst philosophers and scientists (less so in the public at

large)—widespread resistance to or reservations about attributions of morally robust mentality to members of most, if not all, other species" (Acampora 2006, 4). He counters that this cognitive bias rooted in the Cartesian dualism that separates mind from body and that forms the basis for individualism is not, however, borne out in fact: "Where we begin, quite on the contrary, is always already caught up in the experience of being a live body thoroughly involved in a plethora of ecological and social interrelationships with other living bodies and people" (Acampora 2006, 5). This phenomenological, lived body that humans and other animals share forms the basis for his approach to ethics that is rooted in compassion.

2   Compassion has been linked to power and inequality according to a full range of theoretical frameworks. Hannah Arendt argued that compassion creates inequality (see Newcomb 2007). Mary Douglas proclaimed that compassion is repression (see Newcomb 2007). Charles H. Cooley asserted that empathy is not instrumentalized for the purpose of social power, but is a form of power (see Ruiz-Junco 2017).

Importantly for thinking about more-than-human humanitarianism, Arendt had little time for the nonhuman animal, and even scorned and belittled the links between humanitarianism and antivivisection (Cubukcu 2017). Compassion is not language but behavior for Arendt; it is a reactive emotion that she disdains as private, prepolitical, selfish, and based on needs (Newcomb 2007). Here compassion does not distinguish the human but is instead what makes the human an animal, and Arendt is fully committed to maintaining this divide—and with good reason given the links between animalization, dehumanization, and authoritarianism. And so animal studies can learn much from the scholarship that has sought to recuperate compassion for humanitarianism by addressing Arendt's critiques. Compassion has a long history of reproducing inequality (Fassin 2012).

At the same time, we know from a growing social study of emotions that compassion is not simply a biological reflex; like other emotions it is socially structured, patterned, produced, and reproduced (Ruiz-Junco 2017; Hochschild 2016; McCaffree 2020). Compassion is an affective economy (Ahmed 2014, 2004; Berlant 2004a). And as an affective economy, the feeling of compassion is shaped by institutions (e.g., humanitarianism, animal advocacy, science) (see, e.g., Fassin 2012) but also by geographical location and national culture.

For example, Lauren Berlant (2004b, 1) emphasizes in the introduction to an edited volume on compassion that the papers within it were all produced from within the United States, where "the word compassion carries the weight of ongoing debates about the ethics of privilege." It matters that these writings on compassion take place in the United States, and in response to the rhetoric of George W. Bush's "compassionate conservativism." While also about privilege in the United Kingdom—where Tory prime minister David Cameron also played with the idea of "compassionate conservativism" to privilege work but with less religious faith involved—the valence of compassion and ethics is likely differently infected by paternalism and class hierarchy. This would not map onto the United States in a straightforward manner. The affective economies of compassion are heavily politicized, and so place and time matter. Indeed, the idea of compassion being felt by Republicans following Donald Trump's presidency or the Tories under Boris Johnson is unfathomable as naked self-interest and pathological disregard for others has been normalized.

3 Laboratory mice are not rehomed in the ways that dogs and cats are, but a mouse may be rescued by a member of staff who takes the individual animal home.

4 Erica Bornstein develops a relational approach to the closely related emotion of empathy with her concept of "relational empathy." Here she focuses on the ways in which humanitarianism is done in New Dehli in ways that diverge from (and thereby challenge) liberal altruism. Relational empathy works by extending the model of kinship—living with those whom one assists. This is clearly a different relational model from the one proposed here, but both share a focus on relations between people (Bornstein 2012).

5 The life-and-death consequences of an animal technician being replaced were indeed recounted in *The Principles*. Russell and Burch reference Lane-Petter in recounting the following story: "In the same paper, Lane-Petter gave some arresting examples of animal psychosomatics, especially the responses to the behavioral effects of human individuals with whom the animals came into contact. In one guinea pig colony, no deaths had occurred for 5 1/2 months (since it was formed, in fact), until the regular animal technician went on a fortnight's holiday. During the interregnum of another technician, "equally competent and conscientious," four guinea pigs died. Postmortem (including bacteriological) examination gave no clue to the cause of death, and on the return of the original technician the deaths ceased." (Russell and Burch [1959]1967, ch. 6f)

6 I have paid attention to the ways in which Janet's story moved me, inspired by the ways in which Beth Greenhough and Emma Roe (2019) have argued that animal geographies should pay attention to the stories animal technicians tell, as places where "interspecies epiphanies" occur that resist instrumental values and relations. Stories involve specific individuals, human and otherwise, which gets us all out of our roles, if only momentarily, and therein stories also speak to the wider social and political–economic infrastructures that emplace that story. Their work on making storytelling part of the methodological tool kit of animal geographies in laboratories has been crucial to my analysis here.

7 This moment of compassion was marked by the silence that Arendt notes is central to compassion as a feeling. "Passion and compassion are not speechless, but their language consists in gestures and expressions of countenance rather than in words. It is because he listens to the Grand Inquisitor's speech with compassion, and not for lack of arguments, that Jesus remains silent, struck, as it were, by the suffering which lay behind the easy flow of his opponent's great monologue" (Arendt in Newcomb 2007, 112).

8   See also Stevenson (2014, 27–29, 46) on the relationship between animality, seriality, and replaceability in the biopolitical imagination of a humanitarian reason.
9   Robert G. W. Kirk (2018) has shown that *The Principles* languished after its publication but then became central to the 1986 Animals (Science Procedures) Act (ASPA). Here, the 3Rs became the key discourse and practice through which "the good" is pursued in experimental science involving animals.
10 Nathalia Ruiz-Junco (2017) notes that empathy does different things, and one of the things it does is to instantiate instrumental power. This is certainly true of compassion in the context of science that uses laboratory animals, as my own research and that of others has shown (Friese 2013; Giraud 2024; Giraud and Hollin 2016). But the point of this chapter is to show that the instrumentality of compassion can occur alongside what Ruiz-Junco terms the self-transcendent and therapeutic paths of empathy. For a critique of how science enacts instrumentality generally, see Peggs (2013).
11 However, Barbara Prainsack is currently writing a book that develops solidarity in order to address the dangers of climate change. She explores how people are solidaristic with environments here.
12 With affective economy, Ahmed (2014) emphasizes that emotions like compassion do not simply reside in an individual person but rather circulate in ways that connect people. When affective economies, whether involving fear or compassion, do settle down into a specific body, Ahmed emphasizes that processes like racialization occur in tandem with wider discourses.
13 Though it is important to note that Henry Buller (2013) is also critical of the individual-versus-mass dialectic, and notes that individuation of farm animals can be dangerously reductionistic.
14 My use of the first definition has precedence in Donna Haraway's (2008) argument about the ethical importance of suffering with another, which she develops through the story of the scientist who allows themself to be bitten by the mosquitoes that the research animals are being made to be bitten by as part of research. It also has precedence in Astrid Schrader's (2015) work on abyssal intimacy.
15 This definition of compassion overlaps with some elements of the definitions of both empathy and sympathy. Sympathy is generally understood as an emotion in which one sees another suffering and helps them. Sympathy is thus linked with negative emotions and is aligned with pity. Empathy, on the other hand, is understood as an ability to imagine how another feels, which includes both positive and negative emotions and is thus not linked to either helping another or pity. In her interactionist-based proposal for a sociology of empathy, Natalia Ruiz-Junco (2017) shows how Charles H. Cooley's work on sympathy would, however, fall under the word "empathy" according to these definitions.

6. CONSENT

1   Alexandra Palmer, Beth Greenhough, Pru Hobson-West, Gail Davies, and Reuben Message (2023) have explored how scientists use the vernacular of "volunteer-

ing" to describe animal involvement in research. However, this does not generally imply "free, unconstrained, and unpaid" participation but rather a lack of physical restraint. Like consent, an animal truly volunteering is impossible when understanding how animals are positioned in laboratory research. But they argue that this discourse does represent a desire to promote animal welfare by scientists.

2 I am not alone in bringing questions of animal welfare and human research ethics together so that they may speak to one another (e.g., Ashall, Millar, and Hobson-West 2018; Ashall and Hobson-West 2017; Greenhough and Roe 2011).

3 Andrew Fenton (2014) similarly asks if a chimp can say no, but with a different set of theoretical tools and substantive concerns.

4 Within the medicine, informed consent is the vernacular through which the ethics of biomedical research involving humans has been discussed and enforced since the mid-20th century. There are many critiques that address the limits of informed consent as a practice. First, informed consent relies upon the notion of the autonomous individual, and from a relational perspective there are questions about the extent to which certain actors can be understood as autonomous. This is why medicine and medical research often need to go beyond individual consent to include the consent of family members and communities. Second, its practices have been bureaucratized in such a way that may protect research subjects from egregious medical practices but that risk effacing more everyday ethical issues and concerns. In turn, informed consent often protects institutions just as much if not more than it protects people.

5 Paternalism, too, raises the problem of communication not only across different languages, but also across different experiences of the world that make communication within a shared language difficult. In this context, Barnett notes that the giver of humanitarian aid can all too easily slip into the belief that she knows what the receiver wants and needs. Maintaining and sustaining life can mean that consent quite simply needs to be assumed, particularly in emergency situations. For example, Lisa Stevenson's (2014) ethnographic and historical research has shown how a paternalistic ethos, combined with a presumption of consent in humanitarian state building, has resulted in unbearable suffering over generations among Inuit people in Canada. Exploring how tuberculosis and suicide have been addressed as a biopolitical (and thus population-level) problem by the Canadian state, Stevenson shows how the resulting anonymous care creates a devastating indifference to the individual person and their relations. This has left generations of Inuit people listening for traces of the lives and deaths of loved ones. Language can provide a veneer of consent, which can then paper over all the things left unsaid. Stevenson's ethnography forces humanitarians to ask themselves if they are asking the receivers of aid the right questions in seeking consent to their interventions.

6 It is very common for people to talk to the animals they are handling. I have regularly talked to mice, rats, cats, and rhinos encountered in my various research projects.

7 I was unable to see that the mouse was about "to fit." Sarah's attention was so rapid (and indeed she stopped the mouse from fitting by acting quickly) that I could not see this mouse as any different from the others. Sarah's level of knowledge and care was such that I cannot say exactly what "about to fit" looked like.

8 In their analysis of consent in the context of combat sport, Channon and Matthews (2022) find that, even among humans, symbolic consent in the form of spoken and written language is relatively rare and is seen largely to legally protect the organizers from liability and to facilitate coaching. In this sense, formal consent in the context of sport is not dissimilar from informed consent in medical and social research ethics; the language of consent largely protects institutions from legal liabilities (Corrigan 2003). Chanon and Matthews (2022) instead find that ritualized gestures alongside eye contact and a nod of the head form the most common way in which people express their readiness and willingness to fight. But they also find that there is a metalinguistic element to sparring, where partners sense one another's well-being. The danger is that this metalinguistic communication of consent can turn into an assumption of consent.

9 It is important to note here that Peirce, and the development of symbolic interactionist thought that built upon his work, was also challenging a stimulus–response model. This model was developed with dogs (e.g., Pavlov's dog and the Pavlovian response) and then extended to humans. Symbolic interactionism added "interpretation" to this model, arguing that humans do not simply respond to a stimulus but rather interpret that stimulus, based on previous interactions, and act accordingly. What Kohn is doing is extending this "interpretation" stage beyond humans to all of life. This resonates with Hearne's work.

10 And here we can see how interactional communicative dynamics that require metalinguistic capabilities come into dynamic interaction with more structural concerns about the manufacturing of consent in the workplace (Burawoy 1979). Consent here is concerned with the process through which people become complicit in their own oppression, such that consent becomes another mode of coercion through the construction of selves in the Foucaultian sense (see McCabe 2011). I think the important point is that consent and coercion do partially connect, but consent should not be reduced to coercion. The play of mice during a cage cleaning does not have to happen as Sarah enacts this play with the mice. It may benefit the animal facility and the scientists, to be sure. We certainly can see these interactional dynamics as connected to structural process, creating the conditions of possibility for play and its co-optation. But the interactional is not some kind of nested version of these structural dynamics (Tsing 2012), as interactions also contain their own logics and practices that are other than the structural. This is why the interactional is thus a site of potentiality, not only of the potentiality for social reproduction but also for social change.

11 See Bear (2011) on how a particular octopus in an aquarium exists as an individual but also stands as a species representative, and how death works as a particular moment for reasserting the species status of octopi who live in aquariums.

12 See also Brad Bolman (2018, 245) on how jokes offer a means of "rendering laboratory necropolitics manageable, quotidian work. Humour represented a critical method for LEHR researchers to make sense of the contradictory relationships of care that their work required."
13 This is David Graeber's ([2015]2016) argument, developed from an anarchist politics. It is also something that Tarquin Holmes and I (Holmes and Friese 2020) argued in the specific case of the regulation of laboratory animal use by the British states.

CONCLUSION

1 This is, after all, the point of ethnomethodology. And ethnomethodology is the theory–method package that Lynch used to arrive at his analysis of sacrifice.
2 I am inspired here by Joanna Latimer's (2019) work on science under siege.
3 Greenhough and Roe offer an important critique of the deadening consequences when one mode of caring or relating is presented as one among many, particularly given the ways in which animal technicians tell stories of feeling threatened and marginalized. Their care work requires special consideration in this milieu (Greenhough and Roe 2019).
4 Natalie Nuyts and I (Nuyts and Friese 2023) have discussed this, focusing on the 2013 event in which the animal rights organization British Union for the Abolition of Vivisection (BUAV) published "Licensed to Kill," a video created by an undercover BUAV member who gained employment at an animal facility at Imperial College London (ICL). The video presented explicit images of mice being decapitated with a guillotine as well as mice waking up from anaesthesia during experimental procedures. The video led to allegations of incompetence and neglect on the part of scientists at ICL, and sent shock waves through both the British public and scientific community. The Home Office immediately responded with an official investigation (Home Office 2014), while ICL requested an independent investigation (Brown 2013). The government's Animals in Science Committee (ASC) (2014) wrote a report and made recommendations based upon these two investigations.

Across these three reports, several unsatisfactory practices were described, showing how animal rights activism creates change. For example, a significant proportion of the scientists were found to be unaware of the responsibilities attached to their licenses from the Home Office, which are required to conduct animal experiments. Many scientists did not keep adequate records, did not comply with humane end points, and did not report to the Home Office when these humane end points were exceeded. There was an unwillingness to adopt the principle of the 3Rs (replace, reduce, and refine animals in research) in practice, as is required by law. There was insufficient staffing of the ICL animal facility, and communication between scientists and animal-facility staff was deemed substandard. The "culture of care" was labeled poor by the Home Office report. The Brown Report (2013, 24) in part located this in the organizational hierarchy that puts research staff in a position of power toward animal care staff, stating: "The existence of a 'them and us' view (taken of each other by both animal care staff and research colleagues) also did not appear to contribute to the recognition of the expertise of the animal care staff. Nor would this improve the confidence of animal care staff so that they felt able to challenge research staff on animal care and welfare issues."

5   It is worth noting here that we had hypothesized that position in the field of science would be associated with attitudes about the importance of animal care for scientific knowledge (Friese, Nuyts, and Pardo-Guerra 2019). Using multiple correspondence analysis, we described the use of animals in the field of science in the UK as rooted in distinctions between: (1) academic scientists versus nonacademic or industry scientists; (2) those with high versus low cultural and economic capital; and (3) between scientists with a lower status in a higher-esteem institution (e.g., senior technicians and PhD students in or near Oxford, Cambridge, or London) and people with a higher status in a lower-esteem institution (e.g., faculty outside of London and "Oxbridge"). Once the meaning of the axes was established, attitudes about the importance of animal care in scientific research were included as supplementary variables in order to test this hypothesis. There was slight indication that industry scientists place slightly greater importance on animal care, but this was only marginally statistically significant and so we argued that attitudes about animal care were not associated with position in the field. Many scientists in academia and industry have told me that they nonetheless think the survey picked up on a possible truth. There was the belief that industry did care more about animal care because they were experiencing firsthand the confounding influence of poor animal care. And industry scientists were in my experience active in promoting a culture of care.

BIBLIOGRAPHY

Acampora, Ralph R. 2006. *Corporal Compassion: Animal Ethics and Philosophy of Body*. Pittsburgh: University of Pittsburgh Press.
Adams, Carol J. [2006]2007. "The War on Compassion." In *The Feminist Care Tradition*, edited by Joesphine Donovan and Carol J. Adams, 21–36. New York: Columbia University Press.
Adams, Vincanne. 2013. "A History of International Health Encounters: Diplomacy in Transition." In *21st Century Global Health Diplomacy*, edited by Thomas E. Novotny, IIona Kichbusch, and Michaela Told, 41–63. London: World Scientific.
Agamben, Giorgio. 1998. *Homo Sacer: Sovereign Power and Bare Life*. Stanford, CA: Stanford University Press.
Ahmed, Sara. 2004. "Affective Economies." *Social Text 79* 22 (2):117–39.
———. 2014. *The Cultural Politics of Emotion*. Edinburgh: Edinburgh University Press.
Anderson, Alistair, and Pru Hobson-West. 2024. "(Dis)placing Veterinary Medicine: Veterinary Borderlands in Laboratory Animal Research." In *Researching Animal Research. What the Humanities and Social Sciences Can Contribute to Laboratory Animal Science and Welfare*, edited by Gail Davies, Beth Greenhough, Pru Hobson-West, Robert G. W. Kirk, Alexandra Palmer, and Emma Roe, 223–46. Manchester: Manchester University Press.
Animals in Science Committee. 2014. "Lessons to Be Learnt, for Duty Holders and the Regulator, from Reviews and Investigations into Non-Compliance." Last modified July 29, 2019.
Ankeny, Rachel A. 2007. "Wormy Logic: Model Organisms as Case-Based Reasoning." In *Science without Laws: Model Systems, Cases, Exemplary Narratives*, edited by Angela N. H. Creager, Elizabeth Lunbeck, and M. Norton Wise, 46–58. Durham, NC: Duke University Press.
Ankeny, Rachel A., and Sabina Leonelli. 2011. "What's So Special about Model Organisms?" *Studies in History and Philosophy of Science* 42:313–23.
Ankeny, Rachel A., Sabina Leonelli, Nicole C. Nelson, and Edmund Ramsden. 2014. "Making Organisms Model Human Behavior: Situated Models in North-American Alcohol Research, since 1950." *Science in Context* 27 (3):485–509.
Arluke, Arnold. 1991. "Going into the Close with Science: Information Control among Animal Experimenters." *Journal of Contemporary Ethnography* 20 (3):306–30.
Asdal, Kristin. 2018. "'Interested Methods' and 'Versions of Pragmatism.'" *Science, Technology, & Human Values* 43 (4):748–55.

Ashall, Vanessa, and Pru Hobson-West. 2017. "'Doing Good by Proxy': Human-Animal Kinship and the 'Donation' of Canine Blood." *Sociology of Health & Illness* 39 (6):908–22.

Ashall, Vanessa, Kate M. Millar, and Pru Hobson-West. 2018. "Informed Consent in Veterinary Medicine: Ethical Implications for the Profession and the Animal 'Patient.'" *Food Ethics* 1:247–58.

Atanasoski, Neda, and Kalindi Vora. 2019. *Surrogate Humanity: Race, Robots, and the Politics of Technological Futures*. Durham, NC: Duke University Press.

Atkinson, Sarah, Victoria Lawson, and Janine Wiles. 2011. "Care of the Body: Spaces of Practice." *Social & Cultural Geography* 12 (6):563–72.

Atkinson-Graham, Melissa, Martha Kenney, Kelly Ladd, Cameron Michael Murray, and Emily Astra-Jean Simmonds. 2015. "Care in Context: Becoming an STS Researcher." *Social Studies of Science* 45 (5):738–48.

Atwood-Harvey, Dana. 2005. "Death or Declaw: Dealing with Moral Ambiguity in a Veterinary Hospital." *Society & Animals* 13 (4):315–42.

Barad, Karen. 2007. *Meeting the Universe Halfway: Quantum Physics and the Entanglement of Matter and Meaning*. Durham, NC: Duke University Press.

Barnett, Michael. 2011. *Empire of Humanity: A History of Humanitarianism*. Ithaca, NY: Cornell University Press.

Barona, Eduardo, Olga Campos, and Pedro Tomé Martin. 2024. "The 'One Humanitarianism' Approach: Revisiting a Non-Discriminatory Humanitarianism." *Society & Animals*. Published online before print, http://hdl.handle.net/10261/359677.

Bateson, Gregoy. [1972]2000. *Steps to an Ecology of Mind*. Chicago: University of Chicago Press.

Bear, Christopher. 2011. "Being Angelica? Exploring Individual Animal Geographies." *AREA* 43 (3):297–304.

Benedict, Susan. 2003. "Killing while Caring: The Nurses of Hadamar." *Issues in Mental Health Nursing* 24 (1):59–79.

Bennett, Joshua. 2020. *Being Property Once Myself: Blackness and the End of Man*. Cambridge, MA: Belknap Press of Harvard University Press.

Bentham, Jeremy. 1789. *An Introduction to the Principles of Morals and Legislation*. www.earlymoderntexts.com/assets/pdfs/bentham1780.pdf.

Berlant, Lauren. 2004a. *Compassion: The Culture and Politics of an Emotion*. London: Routledge.

———. 2004b. "Introduction." In *Compassion: The Culture and Politics of an Emotion*, edited by Lauren Berlant, 1–13. London: Routledge.

Beumer, Koen. 2014. "Catching the Rat: Understanding Multiple and Contradictory Human-Rat Relations as Situated Practices." *Society & Animals* 22 (1):8–25. doi: https://doi.org/10.1163/15685306-12341316.

Birke, Lynda. 2003. "Who—or What—Are the Rats (and Mice) in the Laboratory." *Society & Animals* 11(3):207–24.

Birke, Lynda, Arnold Arluke, and Mike Michael. 2007. *The Sacrifice: How Scientific Experiments Transform Animals and People*. West Lafayette, IN: Purdue University Press.

Bjorkdahl, Kristian, and Tone Druglitrø. 2016. "Animal Housing/Housing Animals: Nodes of Politics, Practices and Human-Animal Relations." In *Animal Housing and Human-Animal Relations: Politics, Practices and Infrastructures*, edited by Kristian Bjorkdahl and Tone Druglitrø, 1–14. London: Routledge.

Bloor, David. 1991. *Knowledge and Social Imagery*. 2nd ed. Chicago: University of Chicago Press.

Boddice, R. (2016). *The Science of Sympathy: Morality, Evolution, and Victorian Civilization*. Champaign: University of Illinois Press.

Bolker, Jessica A. 2009. "Exemplary and Surrogate Models: Two Modes of Representation in Biology." *Perspectives in Biology and Medicine* 52 (4):485–99.

Bolman, Brad. 2018. "How Experiments Age: Gerontology, Beagles, and Species Projection at Davis." *Social Studies of Science* 48 (2):232–58.

Boris, Eileen, ed. 2010. *Intimate Labors: Cultures, Technologies, and the Politics of Care*. Stanford, CA: Stanford University Press.

Bornstein, Erica. 2012. *Disquieting Gifts: Humanitarianism in New Dehli*. Stanford, CA: Stanford University Press.

Bourdieu, Pierre. 1984. *Distinction: A Social Critique of the Judgment of Taste*. Translated by Richard Nice. Cambridge, MA: Harvard University Press.

———. 1987. *Homo Academicus*. Translated by Peter Collier. Cambridge: Polity.

———. 2006. *Science of Science and Reflexivity*. Cambridge: Polity.

Bowker, Geoggrey C., and Susan Leigh Star. 1999. *Sorting Things Out: Classification and Its Consequences*. Cambridge, MA: MIT Press.

Briggs, Charles L. 2007. "Mediating Infanticide: Theorizing Relations between Narrative and Violence." *Cultural Anthropology* 22 (3):315–56.

Briggs, Laura. 2002. *Reproducing Empire: Race, Sex, Science, and U.S. Imperialism in Puerto Rico*. Berkeley: University of California Press.

Brown, Steven. 2013. "Brown Report: Independent Investigation into Animal Research at Imperial College London." Last modified July 29, 2019.

Bubeck, Marc J. 2023. "Justifying Euthanasia: A Qualitative Study of Veterinarians' Ethical Boundary Work of 'Good' Killing." *Animals* 13 (15):2015. doi: https://doi.org/10.3390/ani13152515.

Buller, Henry. 2013. "Individuation, the Mass and Farm Animals." *Theory, Culture & Society* 30 (7/8):155–75.

Burawoy, Michael. 1979. *The Manufacture of Consent*. Chicago: University of Chicago Press.

Burt, Jonathan. 2006. *Rat*. London: Reaktion Books.

Carbone, Larry. 2024. "Outsiders on the Inside: Citizens and Scholars in Animal Research." In *Researching Animal Research: What the Humanities and Social Sciences Can Contribute to Laboratory Animal Science and Welfare*, edited by Gail Davies,

Beth Greenhough, Pru Hobson-West, Robert G. W. Kirk, Alexandra Palmer, and Emma Roe, 313–18. Manchester: Manchester University Press.

Carroll, Georgie. 2014. *Mouse*. London: Reaktion Books.

Carroll, Lewis. [1865]1993. *Alice's Adventures in Wonderland*. Mineola, NY: Dover Publications.

Casper, Monica J., and Lisa Jean Moore. 2009. *Missing Bodies: The Politics of Visibility*. New York: New York University Press.

Chakrabarti, Pratik. 2012. *Bacteriology in British India: Laboratory Medicine and the Tropics*. Rochester, NY: University of Rochester Press.

Channon, Alex, and Christopher R. Matthews. 2022. "Communicating Consent in Sport: A Typological Model of Athletes' Consent Practices within Combat Sports." *International Review for the Sociology of Sport* 57 (6):899–917.

Charmaz, Kathy. 2014. *Constructing Grounded Theory*. 2nd ed. London: Sage.

Chemla, Karine, and Evelyn Fox Keller, eds. 2017. *Cultures without Culturalism: The Making of Scientific Knowledge*. Durham, NC: Duke University Press.

Chiapperino, Luca. 2021. "Environmental Enrichment: An Experiment in Biosocial Intervention." *BioSocieties* 16 (1):41–69.

Ciobanu, Calina. 2012. "Coetzee's Posthumanist Ethics." *MFS: Modern Fiction Studies* 58 (4):668–98.

Clark, Jonathan L. 2014a. "Labourers or Lab Tools? Rethinking the Role of Lab Animals." In *The Rise of Critical Animal Studies: From the Margins to the Centre*, edited by Nik Taylor and Richard Twine, 139–64. London: Routledge.

———. 2014b. "Living with Transgenic Animals." *Humanimalia* 6 (1):88–113.

Clark, Timothy. 1993. "By Heart: A Reading of Derrida's 'Checosè la poesia?' through Keats and Celan." *Oxford Literary Review* 15 (1/2):43–78.

Clarke, Adele E. 2005. *Situational Analysis: Grounded Theory after the Postmodern Turn*. Thousand Oaks, CA: Sage.

Clarke, Adele E., and Theresa Montini. 1993. "The Many Faces of RU486: Tales of Situated Knowledges and Technological Contestations." *Science, Technology, & Human Values* 18 (1):42–78.

Clarke, Adele E., and Susan Leigh Star. 2008. "The Social Worlds Framework: A Theory/Methods Package." In *The Handbook of Science and Technology Studies*, edited by E. J. Hacket, O. Amsterdamska, M. E. Lynch, and J. Wajcman, 113–37. Cambridge, MA: MIT Press.

Clause, Bonnie Tocher. 1993. "The Wistar Rat as a Right Choice: Establishing Mammalian Standards and the Idea of a Standardized Mammal." *Journal of the History of Biology* 26 (2):329–49.

Cochrane, Alasdair. 2012. *Animal Rights without Liberation: Applied Ethics and Human Obligations*. New York: Columbia University Press.

Coetzee, J. M. [1999]2010. *Disgrace*. London: Vintage Books.

Collective, The Care, Andrea Chatzidakis, Jamie Hakim, Jo Littler, Catherine Rottenberg, and Lynne Segal. 2020. *The Care Manifesto: The Politics of Interdependence*. London: Verso.

Collingwood, Stuart Dodgson. 1898. *The Life and Letters of Lewis Carroll (Rev. C. L. Dodgson)*. London: T. F. Unwin.

Corrigan, Oonagh. 2003. "Empty Ethics: The Problem with Informed Consent." *Sociology of Health & Illness* 25 (3):768–92.

Creager, Angela N. H. 2002. *The Life of a Virus: Tobacco Mosaic Virus as an Experimental Model, 1930–1965*. Chicago: University of Chicago Press.

Çubukçu, Ayça. 2017. "Thinking against Humanity." *London Review of International Law* 5 (2):251–67.

Dam, Mie S., Per T. Sangild, and Mette N. Svendsen. 2018. "Translational Neonatology Research: Transformative Encounters across Species and Disciplines." *HPLS: History and Philosophy of the Life Sciences* 40 (21): https://doi.org/10.1007/s40656-018-0185-2

———. 2020. "Plastic Pigs and Public Secrets in Translational Neonatology in Denmark." *Palgrave Communications* 6 (84): https://doi.org/10.1057/s41599-020-0463-y.

Dam, Mie S., and Mette N. Svendsen. 2018. "Treating Pigs: Balancing Standardisation and Individual Treatments in Translational Neonatology Research." *BioSocieties* 13:349–67.

Daston, Lorraine. 1995. "The Moral Economy of Science." *Osiris* 10:2–24.

Davies, Gail F. 2010. "Captivating Behaviour: Mouse Models, Experimental Genetics and Reductionist Returns in the Neurosciences." *Sociological Review* 58:53–72.

Davies, Gail F. 2011. "Playing Dice with Mice: Building Experimental Futures in Singapore." *New Genetics and Society* 30 (4):433–41.

———. 2012a. "Caring for the Multiple and the Multitude: Assembling Animal Welfare and Enabling Ethical Critique." *Environment and Planning D: Society and Space* 30:623–38.

———. 2012b. "What Is a Humanized Mouse? Remaking the Species and Spaces of Translational Medicine." *Body & Society* 18:126–55.

———. 2013a. "Arguably Big Biology: Sociology, Spatiality and the Knockout Mouse Project." *BioSocieties* 4 (8):417–31.

———. 2013b. "Mobilizing Experimental Life: Spaces of Becoming with Mutant Mice." *Theory, Culture & Society* 30:129–53.

———. 2013c. "Writing Biology with Mutant Mice: The Monstrous Potential of Post Genomic Life." *Geoforum* 48:268–78.

———. 2021. "Locating the 'Culture Wars' in Laboratory Animal Research: National Constitutions and Global Competition." *Studies in History and Philosophy of Science Part A* 89:177–87.

Davies, Gail, Beth Greenhough, Pru Hobson-West, and Robert G. W. Kirk. 2018. "Science, Culture, and Care in Laboratory Animal Research: Interdisciplinary Perspectives on the History and Future of the 3Rs." *Science, Technology, & Human Values* 43 (4):603–21.

Davies, Gail, Beth Greenhough, Pru Hobson-West, Robert G. W. Kirk, Alexandra Palmer, and Emma Roe, eds. 2024. *Researching Animal Research. What the Humanities and Social Sciences Can Contribute to Laboratory Animal Science and Welfare*. Manchester: Manchester University Press.

Davies, Sarah R. 2021. "Atmospheres of Science: Experiencing Scientific Mobility." *Social Studies of Science* 51 (2): 214–32.
Dennison, Ngaire, and Anja Petrie. 2020. "Legislative Framework for Animal Research in the UK." *In Practice* 42 (9):488–96.
Derrida, Jacques. 2008. *The Animal That Therefore I Am*. Translated by David Wills. New York: Fordham University Press.
Despret, Vinciane. 2004. "The Body We Care For: Figures of Anthropo-zoo-genesis." *Body & Society* 10 (2):111–34.
———. 2005. "Sheep Do Have Opinions." In *Making Things Public*, edited by Bruno Latour and Peter Weibel, 360–69. Cambridge, MA: MIT Press.
———. 2008. "The Becomings of Subjectivity in Animal Worlds." *Subjectivity* 23:123–39.
———. 2013. "Responding Bodies and Partial Affinities in Human-Animal Worlds." *Theory, Culture & Society* 30 (7/8):51–76.
———. 2016. *What Would Animals Say if We Asked the Right Questions?* Translated by Brett Buchanan. Minneapolis: University of Minnesota Press.
Donovan, Josephine. 2006. "Feminism and the Treatment of Animals: From Care to Dialogue." *Signs* 31 (2):305–29.
Donovan, Josephine, and Carol J. Adams. 2007. "Introduction." In *The Feminist Care Tradition in Animal Ethics*, edited by Josephine Donovan and Carol J. Adams, 1–20. New York: Columbia University Press.
Druglitrø, Tone. 2016. "Care and Tinkering in the Animal House: Conditioning Monkeys for Poliomyelitis Research and Public Health Work." In *Animal Housing and Human-Animal Relations: Politics, Practices and Infrastructures*, edited by Kristian Bjorkdahl and Tone Druglitrø, 151–66. London: Routledge.
———. 2018. "'Skilled Care' and the Making of Good Science." *Science, Technology, & Human Values* 43 (4):649–70.
Druglitrø, Tone, and Kristin Asdal. 2024. "Experimenting with Care and Cod: On Document-Practices, Versions of Care and Fish as the New Experimental Animal." *Social Studies of Science*. Published online before print, doi: 10.1177/03063127231223904.
Ducey, Ariel. 2010. "Technologies of Caring Labor: From Objects to Affect." In *Intimate Labors: Cultures, Technologies, and the Politics of Care*, edited by Rhacel Salazar Parreñas and Eileen Boris, 18–32. Stanford, CA: Stanford University Press.
Edugyan, Esi. 2018. *Washington Black*. London: Serpent's Tail.
Elias, Ann. 2019. *Coral Empire: Underwater Oceans, Colonial Tropics, Visual Modernity*. Durham, NC: Duke University Press.
Epstein, Steven. 2007. *Inclusion: The Politics of Difference in Medical Research*. Chicago: University of Chicago Press.
Estes, C. L., C. Harrington, and D. N. Pellow. 2001. "The Medical Industrial Complex and the Aging Enterprise: A Critical Perspective." In *Social Policy and Aging: A Critical Perspective*, edited by C. L. Estes, 165–85. Thousand Oaks, CA: Sage.
Fassin, Didier. 2005. "Compassion and Repression: The Moral Economy of Immigration Policies in France." *Cultural Anthropology* 20:362–87.
———. 2007. "Humanitarianism as a Politics of Life." *Public Culture* 19 (3):499–520.

———. 2012. *Humanitarian Reason: A Moral History of the Present.* Berkeley: University of California Press.
Federici, Silvia. 1975. *Wages against Housework.* Bristol: Falling Wall Press and the Power of Women Collective.
Fenton, Andrew. 2014. "Can a Chimp Say 'No'?: Reenvisioning Chimpanzee Dissent in Harmful Research." *Cambridge Quarterly of Healthcare Ethics* 23 (2):130–39.
Forclaz, Amalia Ribi. 2015. *Humanitarian Imperialism: The Politics of Anti-Slavery Activism, 1880–1940.* Oxford: Oxford University Press.
Foucault, Michel. [1976] 020. *The History of Sexuality, the Will to Knowledge.* Vol. 1. New York: Penguin Random House.
Fourcade, Marion. 2009. *Economists and Societies: Discipline and Profession in the United States, Britain, and France, 1890s to 1990s.* Princeton, NJ: Princeton University Press.
———. 2016. "Ordinalization: Lewis A. Coser Memorial Award for Theory Agenda Setting 2014." *Sociological Theory* 34 (3):175–95.
Fox, R., and J. Swazey. 1984. "Medical Morality Is Not Bioethics: Medical Ethics in China and the United States." *Perspectives in Biology and Medicine* 27:337–61.
Franklin, Sarah. 2007. *Dolly Mixtures: The Remaking of Genealogy.* Durham, NC: Duke University Press.
———. 2013. *Biological Relatives: IVF, Stem Cells, and the Future of Kinship.* Durham, NC: Duke University Press.
French, Richard D. 1975. *Antivivisection and Medical Science in Victorian Society.* Princeton, NJ: Princeton University Press.
Friese, Carrie. 2013. "Realizing the Potential of Translational Medicine: The Uncanny Emergence of Care as Science." *Current Anthropology* 54 (7): S129-S138.
———. 2016. "Feminist Animal Care." In *Gender: Macmillan Interdisciplinary Handbooks*, edited by Renee Hoogland, 281–95. Macmillan Reference.
———. 2019. "Intimate Entanglements in the Animal House: Caring for and About Mice." *The Sociological Review* 67 (2):287–298.
———. 2024. "The Shadow Bodies of Mice: Invisible Work in Translational Medicine." *Science, Technology, & Human Values.* Published online before print, doi: https://doi.org/10.1177/01622439241276276.
Friese, Carrie, and Adele E. Clarke. 2012. "Transposing Bodies of Knowledge and Technique: Animal Models at Work in the Reproductive Sciences." *Social Studies of Science* 42 (1):31–52.
Friese, Carrie, and Nathalie Nuyts. 2018. "From the Principles to the Animals (Scientific Procedures) Act: A Commentary on How and Why the 3Rs Became Central to Laboratory Animal Governance in the UK." *Science, Technology, & Human Values* 43 (4):742–47.
Friese, Carrie, Nathalie Nuyts, and Juan Pablo Pardo-Guerra. 2019. "Cultures of Care? Animals and Science in Britain." *British Journal of Sociology* 70 (5):2042–69.
Friese, Carrie, and Joanna Latimer. 2019. "Entanglements in Health and Well-Being: Working with Model Organisms in Biomedicine and Bioscience." *Medical Anthropology Quarterly* 33 (1):120–37.

Friese, Carrie, Tarquin Holmes, and Reuben Message. 2023. "Introduction to National Cultures of Animals, Care and Science." *BioSocieties* 18 (4):707–13.
Garber, Marjorie. 2004. "Compassion." In *Compassion: The Culture and Politics of an Emotion*, edited by Lauren Berlant, 15–27. London: Routledge.
Gilligan, Carol. 1982. *In a Different Voice: Psychological Theory and Women's Development*. Cambridge, MA: Harvard University Press.
Giraud, Eva Haifa. 2019. *What Comes after Entanglement? Activism, Anthropocentrism, and an Ethics of Exclusion*. Durham, NC: Duke University Press.
———. 2024. "Incommensurable Care." In *Researching Animal Research: What the Humanities and Social Sciences Can Contribute to Laboratory Animal Science*, edited by Gail Davies, Beth Greenhough, Pru Hobson-West, Robert G. W. Kirk, Alexandra Palmer, and Emma Roe, 203–10. Manchester: Manchester University Press.
Giraud, Eva, and Gregory Hollin. 2016. "Care, Laboratory Beagles and Affective Utopia." *Theory, Culture & Society* 33 (4): 27–49.
Givoni, Michal. 2016. *The Care of the Witness: A Contemporary History of Testimony in Crises*. Edited by Stefan-Ludwig Hoffman and Samuel Moyn. Cambridge: Cambridge University Press.
Gorman, Richard. 2024. "Outside of Regulations, outside of Imaginations: Why Is It Challenging to Care about Horseshoe Crabs?" In *Researching Animal Research: What the Humanities and Social Sciences Can Contribute to Laboratory Animal Science and Welfare*, edited by Gail Davies, Beth Greenhough, Pru Hobson-West, Robert G. W. Kirk, Alexandra Palmer, and Emma Roe, 57–79. Manchester: Manchester University Press.
Gorman, Richard, and Gail Davies. 2020. "When 'Cultures of Care' Meet: Entanglements and Accountabilities at the Intersection of Animal Research and Patient Involvement in the UK." *Social & Cultural Geography*. doi: 10.1080/14649365.2020.1814850.
Graeber, David. [2015]2016. *The Utopia of Rules: On Technology, Stupidity, and the Secret Joys of Bureaucracy*. London: Melville House.
Greenhough, Beth, and Emma Roe. 2011. "Ethics, Space, and Somatic Sensibilities: Comparing Relationships between Scientific Researchers and Their Human and Animal Experimental Subjects." *Environment and Planning D: Society and Space* 29:47–66.
———. 2018. "Exploring the Role of Animal Technologists in Implementing the 3Rs: An Ethnographic Investigation of the UK University Sector." *Science, Technology, & Human Values* 43 (4):694–722.
———. 2019. "Attuning to Laboratory Animals and Telling Stories: Learning Animal Geography Research Skills from Animal Technologists." *EPD: Space and Society* 37:367–84.
Gruen, Lori. [2004]2007. "Empathy and Vegetarian Commitments." In *The Feminist Care Tradition in Animal Ethics*, edited by Josephine Donovan and Carol J. Adams, 333–43. New York: Columbia University Press.

Guerrini, Anita. 2003. *Experimenting with Humans and Animals: From Galen to Animal Rights*. Baltimore: Johns Hopkins University Press.
Gyasi, Yaa. 2020. *Transcendent Kingdom*. London: Viking.
Haraway, Donna J. 1989. *Primate Visions: Gender, Race, and Nature in the World of Modern Science*. New York: Routledge.
———. 1991. "A Cyborg Manifesto: Science, Technology, and Socialist-Feminism in the Late Twentieth Century." In *Simians, Cyborgs, and Women: The Reinvention of Nature*. New York: Routledge.
———. 1997. *Modest_Witness@Second_Millennium.FemaleMan_Meets_OncoMouse. Feminism and Technoscience*. London: Routledge.
———. 2008. *When Species Meet*. Minneapolis: Minnesota University Press.
Haskell, Thomas L. 1985. "Capitalism and the Origins of the Humanitarian Sensibility, Part 1." *American Historical Review* 90 (2):339–61.
Hernández, Isabel Briz. 2024. "Miscellaneous Care: Bridging the In-between of Translational Medicine." *Science, Technology, & Human Values*. Published online before print, doi: 01622439231223787.
Herrick, Margaret. 2016. "The 'Burnt Offering': Confession and Sacrifice in J.M. Coetzee's Disgrace." *Literature and Theology* 30 (1):82–98.
Hobson-West, Pru. 2009. "What Kind of Animal Is the 'Three Rs'?" *ATLA* 37 (Supplement 2):95–99.
Hochschild, Arlie Russell. 2016. *Strangers in Their Own Land: Anger and Mourning on the American Right*. New York: New Press.
Holmberg, Tora. 2008. "A Feeling for the Animal: On Becoming an Experimentalist." *Society & Animals* 16:316–35.
———. 2011. "Mortal Love: Care Practices in Animal Experimentation." *Feminist Theory* 12 (2):147–63.
Holmes, Tarquin. 2021. "Science, Sensitivity and the Sociozoological Scale: Constituting and Complicating the Human-Animal Boundary at the 1875 Commission on Vivisection and Beyond." *Studies in History and Philosophy of Science Part A* 90(C):194–207.
Holmes, Tarquin, and Carrie Friese. 2020. "Making the Anaesthetised Animal into a Boundary Object: An Analysis of the 1875 Royal Commission on Vivisection." *History and Philosophy of the Life Sciences* 42, 50. doi: https://doi.org/10.1007/s40656-020-00344-9.
———. 2023. "Figuring the 'Cynical Scientists' in British Animal Science: The Politics of Visibility." *BioSocieties* 18:780–800.
Home Office, UK. 2014. "Report of ASRU Investigation into Compliance." Last modified July 29, 2019.
———. 2023. Statistics of Scientific Procedures on Living Animals, Great Britain: 2022. National Statistics.
Humphreys, Michael. 2005. "Getting Personal: Reflexivity and Autoethnographic Vignettes." *Qualitative Inquiry* 11:840–60.

Hurst, Jane L., and Rebecca S. West. 2010. "Taming Anxiety in Laboratory Mice." *Nature Methods* 7 (10):825–26.

Ingold, Tim. 2011. *Being Alive: Essays on Movement, Knowledge and Description*. London: Routledge.

Irvine, Leslie, and Jenny R. Vermilya. 2010. "Gender Work in a Feminized Profession: The Case of Veterinary Medicine." *Gender & Society* 24 (1):56–82.

Jasanoff, Sheila. 2005. *Designs on Nature: Science and Democracy in Europe and the United States*. Princeton, NJ: Princeton University Press.

Jerolmack, Colin, and Shamus Khan. 2014. "Talk Is Cheap: Ethnography and the Attitudinal Fallacy." *Sociological Methods & Research* 43 (2):178–209.

Jeske, Melanie. 2024. "Science Estranged: Power and Inequity in Laboratory Life during the COVID-19 Pandemic." *Science, Technology, & Human Values* 49 (2):263–93.

Johnson, Elizabeth R. 2015. "Of Lobsters, Laboratories, and War: Animal Studies and the Temporality of More-than-Human Encounters." *Environment and Planning D: Society and Space* 33:296–313.

Jones, Susan D. 2003. *Valuing Animals: Veterinarians and Their Patients in Modern America*. Baltimore: Johns Hopkins University Press.

Kafka, Franz. 1919. "In the Penal Colony." Franz Kafka Online, accessed April 1, 2023. www.kafka-online.info.

Kaufmann, Sharon. 2006. *And a Time to Die: How American Hospitals Shape the End of Life*. Berkeley: University of California Press.

Kienzler, Hanna. 2019. "Mental Health System Reform in Contexts of Humanitarian Emergencies: Toward a Theory of 'Practice-Based Evidence.'" *Culture, Medicine and Psychiatry* 43:636–62.

——. 2022. "SymptomSpeak: Women's Struggle for History and Health in Kosovo." *Culture, Medicine and Psychiatry* 46:739–60.

Kilkenny, Carol, William J. Browne, Innes C. Cuthill, Michael Emerson, and Douglas G. Altman. 2010. "The ARRIVE Guidelines: Animal Research: Reporting in Vivo Experiments." *PLoS Biology* 8 (6):e1000412.

Kirk, Robert G. W. 2008. "'Wanted—Standard Guinea Pigs': Standardisation and the Experimental Animal Market in Britain ca. 1919–1947." *Studies in the History and Philosophy of Biology and Biomedical Sciences* 39 (3):280–91.

——. 2010. "A Brave New Animal for a Brave New World: The British Laboratory Animals Bureau and the Constitution of International Standards of Laboratory Animal Production and Use, circa 1947–1968." *Isis* 101:62–94.

——. 2012. "Standardization through Mechanization: Germ-Free Life and the Engineering of the Ideal Laboratory Animal." *Technology and Culture* 53 (1):61–93.

——. 2014. "The Invention of the 'Stressed Animal' and the Development of a Science of Animal Welfare, 1947–86." In *Stress, Shock, and Adaptation in the Twentieth Century*, 241–63. Rochester, NY: University of Rochester Press.

——. 2016. "Care in the Cage: Materializing Moral Economies of Animal Care in the Biomedical Sciences, c. 1945–." In *Animal Housing and Human-Animal Relations:*

Politics, Practices and Infrastructures, edited by Kristian Bjorkdahl and Tone Druglitrø, 167–84. London: Routledge.

———. 2018. "Recovering the Principles of Humane Experimental Technique: The 3Rs and the Human Essence of Animal Research." *Science, Technology, & Human Values* 43 (4):622–48.

Klein, H. J., and K. A. Bayne. 2007. "Establishing a Culture of Care, Conscience, and Responsibility: Addressing the Improvement of Scientific Discovery and Animal Welfare through Science-Based Performance Standards." *ILAR Journal* 48 (1):3–11.

Koch, Lene, and Mette N. Svendsen. 2015. "Negotiating Moral Value: A Story of Danish Research Monkeys and Their Humans." *Science, Technology, & Human Values* 40 (3):368–88.

Kohler, R. E. 1994. *Lords of the Fly: Drosophila Genetics and the Experimental Life*. Chicago: University of Chicago Press.

Kohn, Eduardo. 2013. *How Forests Think: Toward an Anthropology beyond the Human*. Berkeley: University of California Press.

Koolmees, P. A. 2000. "Feminization of Veterinary Medicine in the Netherlands, 1925–2000." *Argos* (Utrecht, Netherlands) 23:125–31.

Krause, Monika. 2014. *The Good Project: Humanitarian Relief NGOs and the Fragmentation of Reason*. Chicago: University of Chicago Press.

Latimer, Joanna. 2000. *The Conduct of Care: Understanding Nursing Practice*. Oxford: Blackwell.

———. 2008. "Unsettling Bodies: Frida Kahlo's Portraits and In/dividuality." *Sociological Review* 56 2:46–62.

———. 2011. "Home, Care and Frail Older People: Relational Extension and the Art of Dwelling." In *Perspectives on Care at Home for Older People*, edited by Christine Ceci, Mary Ellen Purkis, and Kristin Björnsdóttir, 35–61. New York: Routledge.

———. 2013a. "Being Alongside: Rethinking Relations amongst Different Kinds." *Theory, Culture & Society* 30 (7/8):77–104.

———. 2013b. *The Gene, the Clinic and the Family: Diagnosing Dysmorphology, Reviving Medical Dominance (Genetics and Society)*. Edited by Ruth Chadwick, John Dupre, David Wield, and Steve Yearley. London: Routledge.

———. 2019. "Science under Siege? Being alongside the Life Sciences, Giving Science Life." *Sociological Review* 67 (2):64–286.

Latimer, Joanna, and Maria Puig de la Bellacasa. 2013. "Re-thinking the Ethical: Everyday Shifts of Care in Biogerontology" In *Ethics, Law and Society*, vol. 5, edited by Nicky Priaulx and Anthony Wrigley, 153–74. London: Routledge.

Latimer, Joanna, and Daniel Lopez Gomez. 2019a. "Intimate Entanglements: Affects, More-than-Human Intimacies and the Politics of Relations in Science and Technology." *Intimate Entanglements: Sociological Review Monograph Series*. Dorchester, UK: Keele University and Sociological Review Foundation, 3–19.

———. 2019b. *Intimate Entanglements: Sociological Review Monograph Series*. Dorchester, UK: Keele University and Sociological Review Foundation.

Law, John. 2010. "Care and Killing: Tensions in Veterinary Practice." In *Care in Practice: On Tinkering in Clinics, Homes and Farms*, edited by Annemarie Mol, Ingunn Moser, and Jeanette Pols, 57–72. Bielefeld and London: Transcript, Verlag, Transaction.
Lederer, Susan E. 1992. "Political Animals: The Shaping of Biomedical Research Literature in Twentieth-Century America." *Isis* 83 (1):61–79.
Lewis, Jamie, Paul Atkinson, Jean Harrington, and Katie Featherstone. 2013. "Representation and Practical Accomplishment in the Laboratory: When Is an Animal Model Good-Enough?" *Sociology* 47 (4):776–92.
Lincoln, Ann E. 2010. "The Shifting Supply of Men and Women to Occupations: Feminization in Veterinary Education." *Social Forces* 88 (5):1969–98.
Liska, Vivian. 2022. "Law and Sacrifice in Kafka and His Readers." *Interdisciplinary Journal for Religion and Transformation in Contemporary Society* 8 (2):256–74.
Lofstedt, Jeanne 2003. "Gender and Veterinary Medicine." *Canadian Veterinary Journal* 44 (7):533–35.
Logan, Cheryl A. 1999. "The Altered Rationale behind the Choice of a Standard Animal in Experimental Psychology: Henry H. Donaldson, Adolf Meyer, and 'the Albino Rat.'" *History of Psychology* 2 (1):3–24.
———. 2001. "'[A]re Norway Rats . . . Things?'* Diversity versus Generality in the Use of Albino Rats in Experiments on Development and Sexuality." *Journal of the History of Biology* 34 (2):287–314.
———. 2002. "Before There Were Standards: The Role of Test Animals in the Production of Empirical Generality in Physiology." *History of the Journal of Biology* 35 (2):329–63.
———. 2005. "The Legacy of Adolf Meyer's Comparative Approach: Worcester Rats and the Strange Birth of the Animal Model." *Integrative Physiological and Behavioral Science* 40 (4):169–81.
Lopez-Gomez, Daniel. 2019. "What if ANT Wouldn't Pursue Agnosticism but Care?" In *The Routledge Companion to Actor-Network Theory*, edited by Anders Blok, Ignacio Farias, and Celia Roberts, 4–13. London: Routledge.
Lowy, Ilana, and Jean-Paul Gaudilliere. 1998. "Disciplining Cancer: Mice and the Practice of Genetic Purity." In *The Invisible Industrialist: Manufacture and the Production of Scientific Knowledge*, edited by Jean-Paul Gaudilliere and Ilana Lowy, 209–47. Basingstoke, UK: Palgrave Macmillan.
Lundblad, Michael. 2013. *The Birth of the Jungle: Animality in Progressive-Era U.S. Literature and Culture*. Oxford: Oxford University Press.
Lynch, Michael. 1989. "Sacrifice and the Transformation of the Animal Body into a Scientific Object: Laboratory Culture and Ritual Practice in the Neurosciences." *Social Studies of Science* 18:265–89.
Martin, Aryn, Natasha Myers, and Ana Viseu. 2015. "The Politics of Care in Technoscience." *Social Studies of Science* 45 (5):625–41.
Mason, Katherine. 2016. *Infectious Change: Reinventing Chinese Public Health after an Epidemic*. Stanford, CA: Stanford University Press.

Mbembe, Achille. 2019. *Necropolitics*. Translated by Steven Corcoran. Durham, NC: Duke University Press.
McCabe, Darren. 2011. "Accounting for Consent: Exploring the Reproduction of the Labour Process." *Sociology* 45 (3):430–46.
McCaffree, Kevin. 2020. "Towards an Integrative Sociological Theory of Empathy." *European Journal of Social Theory* 23 (4):550–70.
McCarthy, Margaret M., and Frederick S. vom Saal. 1985. "The Influence of Reproductive State on Infanticide by Wild Female House Mice (Mus musculus)." *Physiology & Behavior* 35:843–49.
McGoey, Lindsey. 2007. "On the Will to Ignorance in Bureaucracy." *Economy and Society* 36 (2):212–35.
McLeod, Carmen, and Sarah Hartley. 2018. "Responsibility and Laboratory Animal Research Governance." *Science, Technology, & Human Values* 43 (4):723–41.
Merleau-Ponty, Noemie. 2019. "A Hierarch of Deaths: Stem Cells, Animals and Humans Understood by Developmental Biologists." *Science as Culture* 28 (4):492–512.
Message, Reuben. 2023. "Animal Welfare Chauvinism in Brexit Britain: A Genealogy of Care and Control." *BioSocieties* 18:733–54.
Michael, Mike, and Lynda Birke. 1994. "Accounting for Animal Experiments: Identity and Disreputable 'Others.'" *Science, Technology, & Human Values* 19 (2):189–204.
Mohdin, Aamna. 2020. "Jolyon Maugham Will Not Be Prosecuted for Clubbing Fox to Death." *The Guardian*, UK News, March 5.
Mol, Annemarie. 2008. *The Logic of Care: Health and the Problem of Patient Choice*. London: Routledge.
Mol, Annemarie, Ingunn Moser, and Jeanette Pols. 2010. *Care in Practice: On Tinkering in Clinics, Homes and Farms*. Piscataway, NJ: Transaction.
Morrison, Toni. 1991. *Beloved*. New York: Signet.
Motamedi Fraser, Mariam. 2015. *Word: Beyond Language, beyond Image*. London: Rowman & Littlefield.
———. 2019. "Dog Words—or, How to Think without Language." *Intimate Entanglements: Sociological Review Monograph Series*. Dorchester, UK: Keele University and Sociological Review Foundation, 130–46.
Munro, Rolland. 2004. "Punctualising Identity: Time and the Demanding Relation." *Sociology* 38 (2):293–311.
Murphy, Michelle. 2015. "Unsettling Care: Troubling Transnational Itineraries of Care in Feminist Health Practices." *Social Studies of Science* 45 (5):717–37.
Myelnikov, Dmitry. 2024. "A 'Fragile' Consensus? The Origins of the Animals (Scientific Procedures) Act 1986." In *Researching Animal Research: What the Humanities and Social Sciences Can Contribute to Laboratory Animal Science and Welfare*, edited by Gail Davies, Beth Greenhough, Pru Hobson-West, Robert G. W. Kirk, Alexandra Palmer, and Emma Roe, 29–56. Manchester: Manchester University Press.
Narver, Heather Lyons. 2007. "Demographics, Moral Orientation, and Veterinary Shortages in Food Animal and Laboratory Animal Medicine." *Journal of the American Veterinary Medicine Association* 230 (12):1798–1804.

Nelson, Nicole. 2013. "Modeling Mouse, Human, and Discipline: Epistemic Scaffolds in Animal Behavior Genetics." *Social Studies of Science* 43 (1):3–29.

———. 2018. *Model Behavior: Animal Experiments, Complexity, and the Genetics of Psychiatric Disorders*. Chicago: University of Chicago Press.

Newcomb, Matthew J. 2007. "Totalized Compassion: The (Im)possibilities for Acting Out of Compassion in the Rhetoric of Hannah Arendt." *JAC: Journal of Advanced Composition* 27:105–33.

Nussbaum, Martha C. 1995. *Poetic Justice: The Literary Imagination and Public Life*. Boston: Beacon.

———. 2023. *Justice for Animals: Our Collective Responsibility*. New York: Simon & Schuster.

Nuyts, Nathalie, and Carrie Friese. 2023. "Communicative Patterns and Social Networks in Cultures of Care: Discussing the Morality of Laboratory Animals." *Social & Cultural Geography* 24 (1):11–30. doi: 10.1080/14649365.2021.1901976.

Oudshoorn, Nelly. 2011. *Telecare Technologies and the Transformation of Healthcare*. Basingstoke, UK: Palgrave Macmillan.

Palmer, Alexandra, Beth Greenhough, Pru Hobson-West, Gail Davies, and Reuben Message. 2023. "What Do Scientists Mean when They Talk about Research Animals 'Volunteering'"? *Society & Animals*. Published online before print, doi: https://doi.org/10.1163/15685306-bja10139.

Peggs, Kay. 2009. "A Hostile World for Non-Human Animals: Human Identification and the Oppression of Non-Human Animals for Human Good." *Sociology* 43 (1):85–102.

———. 2013. "Transgenic Animals, Biomedical Experiments and 'Progress.'" *Journal of Animal Ethics* 3 (1):41–56.

Peirce, Charles Sanders. [1894]1998. "What Is a Sign?" In *The Essential Peirce: Selected Philosophical Writings*, edited by the Peirce Edition Project, 4–10. Bloomington: Indiana University Press.

Pereira, S., and M. Tettamani. 2005. "Ahimsa and Alternatives: The Concept of the 4th R. The CPCSEA in India." *ALTEX* 22 (1):3–6.

Pilnick, Alison. 2022. *Reconsidering Patient Centred Care: Between Autonomy and Abandonment*. Leeds, UK: Emerald.

Plankey-Videla, Nancy. 2012. "Informed Consent as Process: Problematizing Informed Consent in Organization Ethnographies." *Qualitative Sociology* 35:1–21.

Pols, Jeanette. 2010. "Telecare: What Patients Care About." In *Care in Practice: On Tinkering in Clinics, Homes and Farms*, edited by Annemarie Mol, Ingunn Moser, and Jeanette Pols, 171–93. Bielefeld and London: Transcript, Verlag, Transaction.

Poole, Trevor. 1997. "Happy Animals Make Good Science." *Laboratory Animals* 31 (2):116–24.

Porcher, Jocelyne. 2017. *The Ethics of Animal Labor: A Collaborative Utopia*. London: Palgrave Macmillan.

Prainsack, Barbara. 2006. "Negotiating Life: The Regulation of Embryonic Stem Cell Research and Human Cloning in Israel." *Social Studies of Science* 36 (2):173–205.

———. 2018. "The 'We' in the 'Me': Solidarity and Health Care in the Era of Personalized Medicine." *Science, Technology, & Human Values* 43 (1):21–44.
Prainsack, Barbara, and Alena Buyx. 2017. *Solidarity in Biomedicine and Beyond*. Cambridge: Cambridge University Press.
Puig de la Bellacasa, Maria. 2011. "Matters of Care in Technoscience: Assembling Neglected Things." *Social Studies of Science* 41 (1):85–106.
———. 2015. "Making Time for Soil: Technoscientific Futurity and the Pace of Care." *Social Studies of Science* 45 (5):691–716.
———. 2017. *Matters of Care: Speculative Ethics in More than Human Worlds*. Minneapolis: University of Minnesota Press.
Rabinow, Paul. 1996. *Essays on the Anthropology of Reason*. Princeton: Princeton University Press.
Rader, Karen. 2004. *Making Mice: Standardizing Animals for American Biomedical Research, 1900–1955*. Princeton: Princeton University Press.
Raffles, Hugh. 2002. "Intimate Knowledge." *International Social Science Journal* 54 (173):325–335.
Rayner, Steve. 2012. "Uncomfortable Knowledge: The Social Construction of Ignorance in Science and Environmental Policy Discourses." *Economy and Society* 41 (1):107–125.
Reardon, J. (2001). The Human Genome Diversity Project: A Case Study in Coproduction. *Social Studies of Science* 31 (3):357–88.
Rheinberger, Hans-Jorg. 2010. *An Epistemology of the Concrete: Twentieth-Century Histories of Life*. Durham: Duke University Press.
Rieff, David. 2002. *A Bed for the Night: Humanitarianism in Crisis*. London: Vintage.
Ritvo, Harriet. 1987. *The Animal Estate: The English and Other Creatures in the Victorian Age*. Cambridge, MA: Harvard University Press.
Rock, Melanie J., and Christopher Degeling. 2015. "Public Health Ethics and More-Than-Human Solidarity." *Social Science & Medicine* 129:61–67.
Roe, Emma, and Beth Greenhough. 2023. "A Good life? A Good Death? Reconciling Care and Harm in Animal Research." *Social & Cultural Geography* 24 (1):49–66.
Rothfels, Nigel. 2008. *Savages and Beasts: The Birth of the Modern Zoo*. Baltimore: Johns Hopkins University Press.
Royal Commission on the Practice of Subjecting Live Animals to Experiments for Scientific Purposes. 1876. *Report of the Royal Commission on the Practice of Subjecting Live Animals to Experiments for Scientific Purposes; with Minutes of Evidence and Appendix*. London: George Edward Eyre and William Spottiswoode.
Royal Society, London. 2014. A Picture of the UK Scientific Workforce. London: Royal Society.
Ruiz-Junco, Nathalia. 2017. "Advancing the Sociology of Empathy: A Proposal." *Symbolic Interaction* 40 (3):414–435.
Russell, William Moy Stratton, and Rex Leonard Burch. [1959]1967. *The Principles of Humane Experimental Technique*. 2nd ed. London: Methuen & Co Ltd.
Sanders, Clinton R. 2003. "Actions Speak Louder than Words: Close Relationships between Humans and Nonhuman Animals." *Symbolic Interaction* 26 (3):405–26.

Savage, Mike. 2021. *The Return of Inequality: Social Change and the Weight of the Past.* Cambridge, MA: Harvard University Press.

Schrader, Astrid. 2015. "Abyssal Intimacies and Temporalities of Care: How (Not) to Care about Deformed Leaf Bugs in the Aftermath of Chernobyl." *Social Studies of Science* 45 (5):665–90.

———. 2017. "Microbial Suicide: Towards a Less Anthropocentric Ontology of Life and Death." *Body & Society* 23 (3):48–74.

Schrader, Astrid, and Elizabeth Johnson. 2017. "Lab Meeting. Considering Killability: Experiments in Unsettling Life and Death." *Catalyst: Feminism, Theory, Technoscience* 3 (2):1–15.

Sewell, Anna. [1877]2014. *Black Beauty.* Scotts Valley, CA: CreateSpace Independent Publishing Platform.

Shapin, Steven, and Simon Schaffer. 1985. *Leviathan and the Air-Pump: Hobbes, Boyle, and the Experimental Life.* Princeton, NJ: Princeton University Press.

Sharp, Lesley A. 2019. *Animal Ethos: The Morality of Human-Animal Encounters in Experimental Lab Science.* Berkeley: University of California Press.

Shaw, Jane R., Brenda N. Bonnett, Debra L. Roter, Cindy L. Adams, and Susan Larson. 2012. "Gender Differences in Veterinarian-Client-Patient Communication in Companion Animal Practice." *Journal of American Veterinary Medicine Association* 241 (1):81–88.

Shmuely, Shira. 2023. *The Bureaucracy of Empathy: Law, Vivisection, and Animal Pain in Late Nineteenth-Century Britain.* Ithaca, NY: Cornell University Press.

Shostak, Sara. 2007. "Translating at Work: Genetically Modified Mouse Models and Molecularization in the Environmental Health Sciences." *Science, Technology, & Human Values* 32 (3):315–38.

Singer, Peter. 1975. *Animal Liberation: A New Ethics for Our Treatment of Animals.* New York: Avon Books.

———. [1975]1995. *Animal Liberation.* London: Random House.

Singleton, Vicky. 2010. "Good Farming: Control or Care?" In *Care in Practice: On Tinkering in Clinics, Homes and Farms*, edited by Annemarie Mol, Ingunn Moser, and Jeannette Pols, 235–56. Bielefeld and London: Transcript, Verlag, Transaction.

Skidmore, Tess. 2024. "'The Place for a Dog Is in the Home': Why Does Species Matter when Rehoming Laboratory Animals?" In *Researching Animal Research: What the Humanities and Social Sciences Can Contribute to Laboratory Animal Science and Welfare*, edited by Gail Davies, Beth Greenhough, Pru Hobson-West, Robert G. W. Kirk, Alexandra Palmer, and Emma Roe, 80–104. Manchester: Manchester University Press.

Slicer, Deborah. 2007. "Your Daughter or Your Dog? A Feminist Assessment of the Animal Research Issue." In *The Feminist Care Tradition in Animal Ethics*, edited by Joesphine Donovan and Carol J. Adams, 105–24. New York: Columbia University Press.

Spiegel, Marjorie. 1996. *The Dreaded Comparison: Human and Animal Slavery.* New York: Mirror Books.

Star, Susan Leigh. 1990. "Power, Technology and the Phenomenology of Conventions: On Being Allergic to Onions." *Sociological Review* 38 (1):26–56.

———. 1993. Cooperation without Consensus in Scientific Problem Solving: Dynamics of Closure in Open Systems. In *Computer Supported Cooperative Work: Cooperation or Conflict?*, edited by S. Esterbrook, 93–105. London: Springer-Verlag.

———. [1988]2015. "The Structure of Ill-Structured Solutions: Boundary Objects and Heterogeneous Distributed Problem Solving." In *Boundary Objects and Beyond: Working with Leigh Star*, edited by Geoffrey C. Bowker, Stefan Timmermans, Adele E. Clarke, and Ellen Balka, 243–59. Cambridge, MA: MIT Press.

Star, Susan Leigh, and James R. Griesemer. 1989. "Institutional Ecology, 'Translations' and Boundary Objects: Amateurs and Professionals in Berkeley's Museum of Vertebrate Zoology, 1907–1939." *Social Studies of Science* 19:387–420.

Star, Susan Leigh, and Anselm Strauss. 1999. "Layers of Silence, Arenas of Voice: The Ecology of Visible and Invisible Work." *Computer Supported Cooperative Work* 8:9–30.

Stevenson, Lisa. 2014. *Life beside Itself: Imagining Care in the Canadian Arctic*. Oakland: University of California Press.

Stowe, Harriet Beecher. [1852]1999. *Uncle Tom's Cabin*. Ware, Hertfordshire, UK: Wordsworth Editions.

Strathern, Marilyn. 1997. "Gender: Division or Comparison?" In *Ideas of Difference*, edited by Kevin Hetherington and Rolland Munro, 42–63. Oxford: Blackwell/Sociological Review.

———. 2004. *Partial Connections*. Updated ed. Walnut Creek, CA: AltaMira.

———. 2020. *Relations: An Anthropological Account*. Durham, NC: Duke University Press.

———. [1992]1995. *After Nature: English Kinship in the Late Twentieth Century*. Cambridge: Cambridge University Press.

Svendsen, Mette N. 2022. *Near Human: Border Zones of Species, Life, and Belonging*. New Brunswick, NJ: Rutgers University Press.

———. 2023. "Pigs, People and Politics: The (Re)drawing of Denmark's Biological, Politico-Geographical, and Genomic 'Borders.'" *BioSocieties* 18 (11):714–32.

Svendsen, Mette N., and Lene Koch. 2013. "Potentializing the Research Piglet in Experimental Neonatal Research." *Current Anthropology* 54 (S7): S118–S128.

Svendsen, Mette N., Iben M. Gjødsbøl, Mie S. Dam, and Laura E. Navne. 2017. "Humanity at the Edge: The Moral Laboratory of Feeding Precarious Lives." *Culture, Medicine and Psychiatry* 41:202–23.

Svendsen, Mette N., Laura E. Navne, Iben M. Gjodsbol, and Mie S. Dam. 2018. "A Life Worth Living: Temporality, Care, and Personhood in the Danish Welfare State." *American Ethnologist* 45 (1):20–33. doi: https://doi.org/10.1111/amet.12596.

Tague, Ingrid H. 2015. *Animal Companions: Pets and Social Change in Eighteenth-Century Britain*. University Park: Pennsylvania State University Press.

Taha, Mai. 2023. "Thinking through the Home: Work, Rent, and the Reproduction of Society." *Social Research: An International Quarterly* 90 (4):837–58.

TallBear, Kim. 2017. "Beyond the Life/Not Life Binary: A Feminist-Indigenous Reading of Cryopreservation, Interspecies Thinking and the New Materialisms." In *Cryopolitics: Frozen Life in a Melting World*, edited by Joanna Radin and Emma Kowal, 179–202. Cambridge, MA: MIT Press.

Tallberg, Linda, and Peter J. Jordan. 2021. "Killing Them 'Softly' (!): Exploring Work Experiences in Care-Based Animal Dirty Work." *Work, Employment and Society*. Published online before print, doi: https://doi.org/10.1177/09500170211008715.

Tavory, Iddo, and Stefan Timmermans. 2014. *Abductive Analysis: Theorizing Qualitative Research*. Chicago: University of Chicago Press.

Thomas, Keith. 1983. *Man and the Natural World: Changing Attitudes in England 1500–1800*. London: Penguin Books.

Thompson, Charis. 2005. *Making Parents: The Ontological Choreography of Reproductive Technologies*. Cambridge, MA: MIT Press.

———. 2013. *Good Science: The Ethical Choreography of Stem Cell Research*. Cambridge, MA: MIT Press.

Thompson, E. P. 1971. "The Moral Economy of the English Crowd in the Eighteenth Century." *Past & Present* 50:76–136.

Thompson, Kirrilly, and Bradley Smith. 2014. "Should We Let Sleeping Dogs Lie ... with Us? Synthesizing the Literature and Setting the Agenda for Research on Human-Animal Co-Sleeping Practices." *Humanimalia* 6 (1). doi: https://doi.org/10.52537/humanimalia.9930.

Traweek, Sharon. 1988. *Life Times and Beamtimes: The World of High Energy Physics*. Cambridge, MA: Harvard University Press.

Tronto, Joan C. 1993. *Moral Boundaries: A Political Argument for an Ethic of Care*. London: Routledge.

Tsing, Anna. 2012. "On Nonscalability: The Living World Is Not Amenable to Precision-Nested Scales." *Common Knowledge* 18 (3):505–24.

———. 2015. *The Mushroom at the End of the World*. Princeton, NJ: Princeton University Press.

Turley, Elliott. 2020. "The Tragicomic Philosophy of *Waiting for Godot*." *Modern Language Quarterly* 81 (3):349–75.

Van der Waal, Rodante. 2024. "The 'Dead Baby Card' and the Early Modern Accusation of Infanticide: Situating Obstetric Violence in the Bio- and Necropolitics of Reproduction." *Feminist Theory*. doi: https://doi.org/10.1177/14647001241245581.

Van der Walt, Johan. 2005. "Interrupting the Myth of the Partage: Reflections on Sovereignty and Sacrifice in the Work of Nancy, Agamben and Derrida." *Law and Critique* 16:277–99.

Vom Saal, Frederick S., Patricia Franks, Michael Boechler, Paola Palanza, and Stefano Parmigiani. 1995. "Nest Defense and Survival of Offspring in Highly Aggressive Wild Canadian Female House Mice." *Physiology & Behavior* 58 (4):669–78.

Vora, Kalindi. 2015. *Life Support: Biocapital and the New History of Outsourced Labor*. Minneapolis: University of Minnesota Press.

Wahlberg, Ayo, and Tine M. Gammeltoft. 2018. *Selective Reproduction in the 21st Century*. Berlin: Springer International.
Wajcman, Judy. 1991. *Feminism Confronts Technology*. Cambridge, UK: Polity.
———. 2004. *Technofeminism*. Cambridge, UK: Polity.
Weber, Elin M., Anna S. Olsson, and Bo Algers. 2007. "High Mortality Rates among Newborn Laboratory Mice: Is It Natural and Which Are the Causes?" *ACTA Veterinaria Scandinavica* 49:S8.
Weil, Kari. 2006. "Killing Them Softly: Animal Death, Linguistic Disability, and the Struggle for Ethics." *Configurations* 14:87–96.
West-McGruer, Kiri. 2020. "There's 'Consent' and Then There's Consent: Mobilising Maori and Indigenous Research Ethics to Problematise the Western Biomedical Model." *Journal of Sociology* 56 (2):184–96.
Wimmer, Andreas, and Nina Glick Schiller. 2002. "Methodological Nationalism and Beyond: Nation-State Building, Migration and the Social Sciences." *Global Networks: A Journal of Transnational Affairs* 2 (4):301–34.
Wolfe, Cary. 2003a. *Animal Rites: American Culture, the Discourse of Species, and Posthumanist Theory*. Chicago: University of Chicago Press.
———. 2003b. *Zoontologies: The Question of the Animal*. Minneapolis: University of Minnesota Press.
———. 2010. *What Is Posthumanism?* Minneapolis: Minnesota University Press.
Wynter, Sylvia, and Katherine McKittrick. 2015. "Unparalleled Catastrophe for Our Species? Or, to Give Humanness a Different Future: Conversations." In *On Being Human as Praxis*, edited by Katherine McKittrick, 9–89. Durham, NC: Duke University Press.
Yair, Gad. 2019. "Hierarchy versus Symmetry in German and Israeli science." *American Journal of Cultural Sociology*. Published online before print, doi: https://doi.org/10.1057/s41290-019-00069-8.

INDEX

Abraham (biblical figure), 82
abyssal intimacy, 176n14
affect, 81
affective economy, 74–75, 81, 89–90, 93, 105, 134, 142–43, 176n12
After Nature (Strathern), 140
Agamben, Giorgio, 90
age, 151
aging, immunity and, 1–2
agricultural animals, 30, 59, 93
Ahmed, Sara, 74–75, 176n12
aid, humanitarian, 109
Alice in Wonderland (Carroll), 118–20
alienation, 41, 52, 169n11
"analytic animal," 76
Anderson, Alistair, 86
anesthesia, 21, 23, 25, 71
anesthetics, 12
animal husbandry, 12–13, 149
animalization, 15, 22, 31, 117
animal justice, 93–94
Animal Research Nexus, 38
"Animal Research: Reporting of *In Vivo* Experiments" (ARRIVE), 13
animal rights, 4, 160n15
Animals (Scientific Procedures) Act (ASPA), 23–24, 30, 176n9
Animals in Science Committee (ASC), 179n4
animal studies, 10, 15–16, 60, 63; animal rights versus, 4; *Disgrace* and, 56; humanitarianism and, 7; languagism in, 124; more-than-human humanitarianism and, 11

animal technicians, 51
The Animal That Therefore I Am (Derrida), 119–20
animal welfare, 8–9, 28, 30, 37, 159n2, 163n3
animal welfare chauvinism, 130
antivivisection, 21–22, 35, 120, 132, 163n4, 174n2
Arendt, Hannah, 93, 117, 174n2, 175n7
Arluke, Arnold, 58
ARRIVE. *See* "Animal Research: Reporting of *In Vivo* Experiments"
ASC. *See* Animals in Science Committee
Asdal, Kristin, 123
ASPA. *See* Animals (Scientific Procedures) Act
assisted reproduction, 77–78
Atanasoski, Neda, 161n23
attachments, 161n20

BALB/c mice, 1–2, 40, 166n2
Barad, Karen, 14, 50
Barnett, Michael, 109
Bateson, Gregory, 108, 112, 121
Beckett, Samuel, 128–29
"becoming alongside," 49
"becoming with," 10
"being alongside," 5, 10, 33–34, 43, 49, 106, 121, 132
Beloved (Morrison), 68–69
Benjamin, Walter, 90
Bennett, Joshua, 15–17, 31, 50
Bentham, Jeremy, 20, 74, 171n1
Berlant, Lauren, 174n2
biography, 138

201

biomedicine, 2, 11, 35, 64, 86, 154; consent and, 123; humanitarian aspects of, 3
biopolitics, 55, 124–25; agriculture and, 80
biosecurity, 76
Birke, Lynda, 58
Black Beauty (Sewell), 20
Bolman, Brad, 89, 159n1
boundary objects, 23, 164n5
Bourdieu, Pierre, 141
Boyle, Robert, 21
Brexit, 127
Britishness, 18, 129–32, 155
British Union for the Abolition of Vivisection (BUAV), 179n4
British Veterinary Association (BVA), 24
Brown Report, 179n4
BUAV. *See* British Union for the Abolition of Vivisection
Buller, Henry, 57, 70–71, 93, 105, 176n13
Burch, Rex Leonard, 9, 30, 97–98, 102, 175n5
Buyx, Alana, 93, 103, 104
BVA. *See* British Veterinary Association

cages, 71
Cameron, David, 174n2
cancer, 1–2, 64
capitalism, 41, 160n13
Carbone, Larry, 37, 38
care, 6; culture of, 13, 90, 141–42, 179n4, 180n5; ethics of, 51–53; haptic, 39, 134–35; health care, 104; humanitarianism through prism and, 53–54; incommensurable, 9, 38; miscellaneous, 166n4; more-than-human humanitarianism and, 50–53; more-than-human knowing and, 45–47; transcendent science and, 47–50; vignette on, 42–44; as work and knowledge, 40–42
care work: animal, 12–13; devaluation of, 40; marginalization of, 19; as unpaid, 41
"caring about," 40–41

caring across distance, 30
"caring for," 40–41
Carroll, Lewis, 118–20
Cartesian mind, 48
Carusi, Anna-Maria, 159n6
Casper, Monica, 3
Centers for Disease Control (CDC), 173n8
Chemla, Karine, 130
choice, 165n1
Christianity, 22, 35
Ciobanu, Calina, 61, 82, 84, 171n8
civic epistemologies, 6
Clark, Timothy, 121
Clarke, Adele, 3, 164n5
class, 31
classism, 23
Cochrane, Alasdair, 160n15
coercion, 112, 117
Coetzee, J. M., 14, 16–17, 59–63, 75, 82–85, 91, 121, 171n8
colonialism, 22, 73, 116–17, 135, 141
Committee for Reform of Animal Experimentation (CRAE), 23–24
"the commons," 87
compassion, 92–93, 117, 135, 174n2, 175n7; humanitarianism through prism and, 105–7; more-than-human humanitarianism and, 102–5; replaceability and, 98–102; replaceability as ethical mandate, 97–98; vignette on, 94–97
compassionate conservativism, 174n2
consent, 108, 112–15, 178n8, 178n10; amplifying, 116–18; ethics and poetics, 120–22; humanitarianism through prism, 124–26; informed, 109, 123, 177n3; more-than-human humanitarianism and, 109, 122–24; vignette on, 110–11, 118–20
Cooley, Charles H., 174n2, 176n15
cost, of animal welfare, 159n2
COVID-19 pandemic, 2, 84, 128, 159n5
CRAE. *See* Committee for Reform of Animal Experimentation

Cruelty to Animals Act (1876), 12, 21, 23, 27, 30, 123, 155, 163n4
culture: of care, 13, 90, 141–42, 179n4, 180n5; "culture of no culture," 128; national, 130; "waiting for culture," 127, 128
Culture of Care Barometer, 142

Dam, Mie, 80, 162n1
Daston, Lorraine, 79, 172n5
Davies, Gail, 5, 24, 36, 133, 141, 166n2, 176n1
death, 56, 77; meaningful way to die, 79–82; normalization of, 58
Degeling, Christopher, 104
dehumanization, 15
dental malocclusion, 29
Derrida, Jacques, 18, 75, 82, 90–91, 119–22
Despret, Vinciane, 50, 124–25
detachments, 161n20
Dickens, Charles, 20
difference, 38
"diffractive reading," 14
discomfort, freedom from, 30
disease: foot-and-mouth, 170n6; freedom from, 30; zoonotic, 104
Disgrace (Coetzee), 14, 16–17, 59–63, 75, 82–85, 91, 121, 171n8
distress: freedom from, 30; mental and social, 27
"dividual," 34
documentary moralities, 72
dog gestures, 113–15. *See also* word relations
dogs, relationships with, 96
Douglas, Mary, 117, 174n2
Druglitrø, Tone, 51, 123, 166n5

Edugyan, Esi, 14, 16–17, 93, 99–102, 116–17
emotions, 74–75, 81, 95
empathy, 105, 174n2, 176n10, 176n15; interspecies, 31, 33; maps, 102, 106, 134; paths, 102, 106; relational, 174n4

Englishness, 137
"epistemic objects," 162n1
ethics, 60, 129; of care, 51–53; consent and, 120–22; Kantian, 55; posthuman, 56, 84; self and, 139; virtue, 15, 123
European Convention on Human Rights, 18
European Union Directive 2010/63/EU, 24
euthanasia, 169n11
experiments, 23, 25–26, 56, 80–81, 103; expanding notion of suffering and, 27–28; physiological, 12; sacrifice and, 75; Slicer on, 59; 3Rs (replace, reduce, and refine) for, 9, 30, 93, 97. *See also* antivivisection; vivisection
extinction, 70

farm animals. *See* agricultural animals
Fassin, Didier, 136, 138–39
fear, freedom from, 30
"feeling alongside," 106
feminism, 58; liberal, 50; Marxist, 166n3; second-wave, 72
feminization, of veterinary medicine, 52
fiction, 14–15, 19–20, 56, 128
Five Freedoms, 30
foster mothers (mice), 69
Foucault, Michel, 55, 178n10
Fourcade, Marion, 142
FRAME. *See* Fund for the Replacement of Animals in Medical Experiments
Fraser, Mariam Motamedi, 110, 113–16, 122, 124–25, 166n7
freedom: Five Freedoms, 30; individual, 123
French, Richard, 21, 163n4
Fund for the Replacement of Animals in Medical Experiments (FRAME), 24

Garber, Marjorie, 106
gender, 31, 151
geriatric mice, 2, 87

Gilligan, Carol, 40, 42, 50, 53
Giraud, Eva Haifa, 9, 16
Givoni, Michal, 139
Good Samaritans, 125
Gorman, Richard, 141, 172n3
governing, 85–88
Greenhough, Beth, 51, 58, 115, 159n2, 166n5, 176n1, 179n3
Guerrini, Anita, 21
Gyasi, Yaa, 14, 16–17, 19–20, 31–35, 47–50, 133, 166n8

haptic care, 39, 134–35
Haraway, Donna, 10–11, 36, 49–50, 57–58, 133, 137, 166n3, 176n14
Hardy, Thomas, 82, 171n8
Harvey, William, 21
Haughton, Samuel, 171n1
health care, 104
Hearne, Vicki, 113
heat, 44, 46–47, 166n7
hedgehogs, 120–22
Heidegger, Martin, 50
Herriot, James, 86
Hobson-West, Pru, 86, 176n1
Hochschild, Arlie, 102, 134
Holmberg, Tora, 58
Holmes, Tarquin, 5, 12, 20, 71, 89, 155, 160n12, 163nn4–5, 171n1
Home Office, 27–28, 43, 72, 153
Hooke, Robert, 21, 27, 35
horseshoe crabs, 172n3
"humane," 98; killing, 23
human exceptionalism, 98, 139
humanism, 6
humanitarianism, 5, 140; of biomedicine, 3; capitalism and, 160n13; caring across distance and, 30; consent and, 109; politicization of, 8, 37; rethinking, 7–9; suffering and, 20–25; through prism, 36–38, 53–54, 71–73, 90–91, 105–7, 124–26, 135–37. *See also* more-than-human humanitarianism

humanitarian wars, 8, 72
human rights, 7–9, 11, 36–38, 90, 135
humor, 119, 127–28, 178n12
hunger, freedom from, 30
hypoxia, 77

IACUC. *See* Institutional Animal Care and Use Committee
ICL. *See* Imperial College London
ICRC. *See* International Committee of the Red Cross
immunity, aging and, 1–2
immunization, 127
Imperial College London (ICL), 179n4
imperialism, 15, 22
implicated actors, 3
incommensurable care, 9, 38
index/indices, 45
In a Different Voice (Gilligan), 42
industrialization, 22
inequality, 75, 87
inequity, 140–43
infanticide, 64–69, 171n10
infection, absence of, 71
informed consent, 109, 123, 177n3
injury, freedom from, 30
"In the Penal Colony" (Kafka), 84, 90
the Institute, 28, 39, 65, 72, 86, 155
Institutional Animal Care and Use Committee (IACUC), 24
institutional change, 13
International Committee of the Red Cross (ICRC), 4, 138
interspecies relations, 16; communication and, 17, 31, 33, 123, 124; empathy and, 31, 33; epiphanies and, 175n6
intimate entanglements, 115
intimate knowledge, 53, 109, 123, 124, 134
invisibility, 3
invisible work, 3
Irvine, Leslie, 168n10
Isaac (biblical figure), 82

Johnson, Boris, 88, 174n2
Jones, Susan, 167n9
Jude the Obscure (Hardy), 82, 171n8
justice, 42, 93–94, 105–6, 168n10
Justice for Animals (Nussbaum), 105
juxtaposition, 14

Kafka, Franz, 84, 90
Kantian ethics, 55
Keller, Evelyn Fox, 130
Kienzler, Hanna, 44, 126
killability, 57–59, 71–72
killing, 17–18, 55–56, 74; humane, 23; humanitarianism through prism, 71–73; infanticide and, 66–69; killability and, 57–59, 71–72; more-than-human humanitarianism and, 69–71; relations, 57, 60–63, 133; vignette on, 64–66
Kingsolver, Barbara, 31
kinship, 10, 67
Kirk, Robert G. W., 27–28, 71, 163n3, 176n9
Klein, Emanuel, 164n5
knowing: more-than-human, 45–47; relational models of, 42, 50
knowledge: care as, 40–42; intimate, 53, 109, 123, 124, 134
Koch, Lene, 79
Kohn, Eduardo, 45–47, 112–13, 116, 122, 124, 133, 172n6, 178n9
Krause, Monika, 125

labor: under capitalism, 41; physical, 111; replaceability and, 101
laboratory animals, 11–13. *See also specific topics*
Lane-Petter, William, 103, 175
language, 46, 177n5
languagism, 124
Latimer, Joanna, 5, 10–11, 33–34, 49–50, 108, 152, 155, 161n20, 170n3
Latour, Bruno, 165n1
Law, John, 170n6

Leviathan and the Air-Pump (Shapin and Schaffer), 137
life sciences, sacrifice in, 75–76, 80
longue durée, 52, 54
love for animals, 22
Lynch, Michael, 76, 80, 128

marginalization: of care work, 19; of relational ways of knowing, 42
Mason, Katherine, 86–87, 173n8
Matthews, Christopher R., 124, 178n8
meaningful way to die, 79–82
Médecins Sans Frontières (MSF), 4, 138
medical breakthroughs, 75
medical-industrial complex, 13
medical treatment, 129
mental distress, 27
Merleau-Ponty, Noemie, 170n3
Message, Reuben, 130, 176n1
metalinguistics, of play, 112
methodological nationalism, 18
mice. *See specific topics*
Michael, Mike, 58
microsociology, 52
miscellaneous care, 166n4
model organisms, 19
modernity, 173n8
Mol, Annemarie, 165n1, 170n6
monkeys, 45, 47
Moore, Lisa Jean, 3
"moral economy," 79, 85, 172n5
moral reasoning, 42
morals, 8, 15, 22–23, 42, 59
more-than-human humanitarianism, 4, 6–10, 16–18, 24, 54–59, 128, 131–35, 155–57; animal studies and, 11; care and, 50–53; compassion and, 102–5; consent and, 109, 122–24; killing and, 69–71; sacrifice and, 89–90; suffering and, 35–36
more-than-human knowing, 45–47
Morrison, Toni, 68–69
mouse information system, 29

MSF. *See* Médecins Sans Frontières
multispecies suffering, 31–35
mustard seeds, 40
Myelnikov, Dmitry, 23

Named Animal Care and Welfare Officers (NACWOs), 47
Named Veterinarian Surgeons, 86
Narver, Heather Lyons, 168n10
national cultures, 130
National Health Services (NHS), 2–3, 142
nationalism, 18, 131
nationality, 130, 151
"naturalistic animal," 76
neck, dislocation of, 77, 136
necropolitics, 55, 73, 134
Nelson, Nicole, 162n1
neutrality, 37
Newcomb, Matthew, 93
NHS. *See* National Health Services
normalization, of death, 58
Nussbaum, Martha, 15, 105
Nuyts, Nathalie, 5–6, 130, 152–53, 179n4

One Humanitarianism, 161n18
oppression, 52
ordinalization, 142
Owens, Delia, 31

pain, 12, 23, 25; freedom from, 30; standardized pain management tools, 37–38
Palmer, Alexandra, 176n1
Pardo-Guerra, Juan Pablo, 5–6, 130, 152
partial connections, 10, 14, 121, 132, 161n18
paternalism, 109, 131, 135, 174n2, 177n5
patriarchy, 41
Peirce, Charles Sanders, 45
PETA, 105
pharmaceutical research, 2
physical labor, 111
physiology, 12, 21

pigs, 80, 81, 82
play, 122; metalinguistics of, 112
Poetic Justice (Nussbaum), 15
poetics, consent and, 120–22
polarization cycle, 4, 138
politicization, of humanitarianism, 8, 37
posthuman ethics, 56, 84
Prainsack, Barbara, 93, 103–4, 176n11
The Principles of Humane Experimental Technique (Russell and Burch), 9, 30, 97–98, 102, 175n5
prism ethnography, 14
psychology, 33–35, 42
PTEN deletion, 64
public health, 85–88, 90, 104
Puig de la Bellacasa, Maria, 41, 165n1

quarantine, 85, 87

racialization, 15, 31, 117–18
racism, 15, 23, 31, 34, 41, 72, 101, 116
rats, 24, 26, 39, 133, 155, 165n7; compassion and, 92; sepsis and, 25; suffering and, 27–28, 30, 35
reattachments, 161n20
Reiff, David, 8
relational models: of becoming, 45; of empathy, 174n4; of knowing, 42, 50
relational ways of knowing, marginalization of, 42
religious doctrine, 6
replace, reduce, and refine (3Rs), 9, 30, 56, 93, 97, 123, 176n9, 179n4
replaceability: compassion and, 98–102; as ethical mandate, 97–98
Researching Animal Research (Carbone), 37
rights: animal, 4, 160n15; human, 7–9, 11, 36–38, 90, 135
Ritvo, Harriet, 22
Rock, Melanie, 104
Roe, Emma, 51, 58, 115, 159n2, 166n5, 179n3
Rothfels, Nigel, 117–18

Royal Commission on Vivisection (1875), 5, 12, 23, 71, 89, 155, 163n4
Royal Society for the Prevention of Cruelty to Animals (RSPCA), 171n11, 173n7
Royal Society for the Protection of Birds (RSPB), 173n7
RSPCA. *See* Royal Society for the Prevention of Cruelty to Animals
Ruiz-Junco, Natalia, 102, 176n10, 176n15
Russell, William Moy Stratton, 9, 30, 97–98, 102, 175n5

sacrifice, 18, 74, 96, 128, 134, 141–42, 172n4; in *Disgrace*, 82–85; governing and, 85–88; humanitarianism through prism and, 90–91; in life sciences, 75–76, 80; as meaningful way to die, 79–82; more-than-human humanitarianism and, 89–90; self-sacrifice, 88; vignette on, 76–79
The Sacrifice (Birke, Arluke, and Michael), 58
sadness, 74–75, 79, 81, 95, 134
salvage, 6
Sanders, Clinton, 96
Saussure, Ferdinand de, 46
Savage, Mike, 141, 142
saviorism, 55, 72
Schaffer, Simon, 137
Schrader, Astrid, 40, 55, 176n14
science and technology studies (STS), 3, 5, 42, 45, 69, 128, 150, 165n1
selective breeding, 162n1
self: ethical, 139; ownership, 123; sacrifice, 88; transcendence of, 50
sepsis, 25
Sewell, Anna, 20
sexism, 42
Shapin, Steven, 137
Sharp, Lesley, 24, 58, 67–68, 76, 89, 96, 119, 171n9
Shmuely, Shira, 163n4

signs, 45, 46
Singer, Peter, 53, 105
situational analysis, 157
slavery, 20, 68–69, 99, 101–2
Slicer, Deborah, 59
social change, 140–43
social reproduction, 166n3
social worlds, 164n5
society, kinship and, 10
sociozoological scale, 170n3
solidarity, 93–94, 102–5, 176n11
song, 98
standardization, 52, 162n1, 166n5; of pain management tools, 37–38
Star, Susan Leigh, 3, 164n5
Stengers, Isabelle, 165n1
Stevenson, Lisa, 98, 124–25, 177n5
stimulus–response model, 113–14
Stowe, Harriet Beecher, 20
Strathern, Marilyn, 10–11, 14, 18, 34, 140, 142
Strauss, Anselm, 3, 151
stress, 27, 28, 163n3
STS. *See* science and technology studies
suffering, 12, 17, 19, 102, 135; alleviating, 132; expanding the notion of, 27–30; humanitarianism and, 20–25; more-than-human humanitarianism, 35–36; multispecies, 31–35; preventing, 30; shared, 26, 36; vignette on, 25–27, 28–30
surveillance work, 29–30
Svendsen, Mette, 8, 14, 79–82, 131, 156, 162n1, 172n4, 172n6
symbolic interactionism, 178n9
sympathy, 53, 176n15
SymptomSpeak, 126

Tague, Ingrid H., 22
TallBear, Kim, 50–51
teeth, of mice, 29
telemetry, 25, 26, 27
Thatcher, Margaret, 23

"A Theory of Play and Fantasy" (Bateson), 108
thirst, freedom from, 30
Thomas, Keith, 6
Thompson, Charis, 162n2
Thompson, E. P., 172n5
3Rs (replace, reduce, and refine), 9, 30, 56, 93, 97, 123, 176n9, 179n4
transcendence, 91
Transcendent Kingdom (Gyasi), 14, 16–17, 19–20, 31–35, 47–50, 133, 166n8
transcendent science, 47–50
transgenic mice, 65–66, 70, 159n4
transgenics, 162n1
translational medicine, 155, 166n4
trauma: dental malocclusion and, 29; psychological model of, 33–34; recovery model of, 33–34
Tronto, Joan, 40
Tsing, Anna, 166n3, 169n11

UFAW. *See* Universities Federation for Animal Welfare
UFAW Handbook on the Care and Management of Laboratory Animals, 97
Uncle Tom's Cabin (Stowe), 20
United States Animal Welfare Act, 24
universal health care, 104
Universities Federation for Animal Welfare (UFAW), 97

vaccines, 2–3, 21, 84, 127
Valuing Animals (Jones), 167n9

Vermilya, Jenny R., 168n10
veterinary medicine, 51–52, 167n9
Victorian Britain, 6, 8, 20, 135, 171n1
violence, 7, 58–59, 68, 99, 161n19; distance and, 38; of racism, 101; sexual, 62
virtue ethics, 15, 123
visibility, 3
vivisection, 12, 21, 89. *See also* antivivisection
Vora, Kalindi, 14, 161n23
vulnerability, 95

van der Waal, Rodante, 171n10
"waiting for culture," 127, 128
Waiting for Godot (Beckett), 128–29
Wajcman, Judy, 52, 166n3
van der Walt, Johan, 75, 91
wars, humanitarian, 8, 72
Washington Black (Edugyan), 14, 16–17, 93, 99–102, 116–17
Weil, Kari, 59, 60, 170n7
welfare, animal, 8–9, 30, 37, 159n2
well-being, animal, 28, 163n3
wellness, 47
witnessing, 4, 137–40
women, 126, 167n9, 168n10; in veterinary medicine, 51–52
word relations, 115
work: care as, 40–42; invisible, 3; surveillance, 29–30. *See also* care work
World War I, 8

zoonotic diseases, 104

ABOUT THE AUTHOR

CARRIE FRIESE is Associate Professor of Sociology at the London School of Economics and Political Science. She is the author of *Cloning Wild Life: Zoos, Captivity, and the Future of Endangered Animals.*

www.ingramcontent.com/pod-product-compliance
Lightning Source LLC
Chambersburg PA
CBHW031149020426
42333CB00013B/582